IARIA

AA004276

GLOBAL HEALTH 2016

The Fifth International Conference on Global Health Challenges

October 9 - 13, 2016

Venice, Italy

GLOBAL HEALTH 2016 Editors

Hassan Khachfe, Lebanese International University, Lebanon

Wolfgang Leister, Norsk Regnesentral, Norway

Printed from e-media with permission by:

Curran Associates, Inc.
57 Morehouse Lane
Red Hook, NY 12571

Some format issues inherent in the e-media version may also appear in this print version.

Printed by Curran Associates, Inc. (2016)

International Academy, Research, and Industry Association (IARIA)
412 Derby Way
Wilmington, DE 19810

Phone: (408) 893-6407
Fax: (408) 527-6351

petre@iaria.org

Additional copies of this publication are available from:

Curran Associates, Inc.
57 Morehouse Lane
Red Hook, NY 12571 USA
Phone: 845-758-0400
Fax: 845-758-2633
Email: curran@proceedings.com
Web: www.proceedings.com

TABLE OF CONTENTS

GLOBAL HEALTH 1: TECHNOLOGY

GLOBAL HEALTH 2: CHALLENGES

GLOBAL HEALTH 3: MEDICAL SYSTEMS AND TECHNOLOGIES I

GLOBAL HEALTH 4: MEDICAL SYSTEMS AND TECHNOLOGIES II

GLOBAL HEALTH 2016

Forward

The Fifth International Conference on Global Health Challenges (GLOBAL HEALTH 2016), held between October 9 and 13, 2016 in Venice, Italy, continued a series of events taking a global perspective on population health, from national to cross-country approaches, multiplatform technologies, from drug design to medicine accessibility, everything under mobile, ubiquitous, and personalized characteristics of new age population.

Recent advances in technology and computational science influenced a large spectrum of branches in approaching population health. Despite significant progresses, many challenges exist, including health informatics, cross-country platforms interoperability, system and laws harmonization, protection of health data, practical solutions, accessibility to health services, and many others. Along with technological progress, personalized medicine, ambient assistance and pervasive health complement patient needs. A combination of classical and information-driven approach is developing now, where diagnosis systems, data protection mechanisms, remote assistance and hospital-processes are converging.

The event was very competitive in its selection process and very well perceived by the international scientific and industrial communities. As such, it has attracted excellent contributions and active participation from all over the world. We were very pleased to receive a large amount of top quality contributions.

The conference had the following tracks:

- Technology
- Challenges
- Medical Systems and Technologies
- Healthcare Economics and Biobanking Innovation

We take here the opportunity to warmly thank all the members of the GLOBAL HEALTH 2016 technical program committee, as well as the numerous reviewers. The creation of such a high quality conference program would not have been possible without their involvement. We also kindly thank all the authors that dedicated much of their time and effort to contribute to GLOBAL HEALTH 2016. We truly believe that, thanks to all these efforts, the final conference program consisted of top quality contributions.

Also, this event could not have been a reality without the support of many individuals, organizations and sponsors. We also gratefully thank the members of the GLOBAL HEALTH 2016 organizing committee for their help in handling the logistics and for their work that made this professional meeting a success.

We hope GLOBAL HEALTH 2016 was a successful international forum for the exchange of ideas and results between academia and industry and to promote further progress in the area of global health challenges.

We also hope that Venice, Italy, provided a pleasant environment during the conference and everyone saved some time to enjoy the unique charm of the city.

GLOBAL HEALTH Advisory Committee

Hassan Khachfe, Lebanese International University, Lebanon
Shunji Shimizu, Tokyo University of Science, Suwa, Japan
Sten Hanke, IPN Instituto Pedro Nunes, Portugal | AIT Austrian Institute of Technology GmbH, Austria
Michael Bremseth, Winthrop University, USA
Jean-Pierre Cordeau, McGill University Health Centre, Canada
Ivan Evgeniev, TU Sofia, Bulgaria
Matthieu-P. Schapranow, Hasso Plattner Institute, Germany
Abdel-Badeeh M. Salem, Ain Shams University Abbasia- Cairo Egypt

GLOBAL HEALTH Technology Advisory Committee

Marike Hettinga, Windesheim University of Applied Sciences, Netherlands
Shirley Coyle, Centre for Sensor Web Technologies / Dublin City University, Ireland
Anna Cysewska-Sobusiak, Poznan University of Technology, Poland
Kevin Daimi, University of Detroit Mercy, USA
Ravi Janardan, University of Minnesota-Twin Cities, USA
Jose M. Juarez, University of Murcia, Spain
Jochen Meyer, OFFIS Institute for Computer Science, Germany
Vesselin Gueorguiev, TU Sofia, Bulgaria
Miki Sirola, Aalto University, Finland
Daniel Steffen, German Research Center for Artificial Intelligence (DFKI) GmbH - Kaiserslautern, Germany
Maria Taboada Iglesias, University of Santiago de Compostela, Spain
Jari Veijalainen, University of Jyvaskyla, Finland
Pierangelo Veltri, University Magna Graecia of Catanzaro, Italy
Brijesh Verma, Central Queensland University, Australia
Liangjiang Wang, Clemson University - Greenwood, USA
Hans C. Ossebaard, National Institute for Public Health and the Environment - Bilthoven, The Netherlands
Ramesh Krishnamurthy, Health Systems and Innovation Cluster, World Health Organization - Geneva, Switzerland

GLOBAL HEALTH Public/Government/Research Advisory Committee

Christopher Mayer, AIT Austrian Institute of Technology GmbH, Austria
João Quintas, Instituto Pedro Nunes, Portugal
Yoshikazu Seki, National Institute of Advanced Industrial Science and Technology (AIST) - Ibaraki, Japan

Hrvoje Belani, Croatian Health Insurance Fund - Zagreb, Croatia
Stefan J. Darmoni, Rouen University Hospital, France
Chih-Lin Chi, University of Minnesota-Twin Cities, USA
Siddhartha Jonnalagadda, Northwestern University, USA
Bin Song, Oracle, Inc., USA
Egon L. van den Broek, Utrecht University - Utrecht, The Netherlands
Zhongming Zhao, Vanderbilt University / Medical Center, USA
Hua Xu, Vanderbilt University/School of Medicine, The Netherlands

GLOBAL HEALTH 2016

Committee

GLOBAL HEALTH Advisory Committee

Hassan Khachfe, Lebanese International University, Lebanon
Shunji Shimizu, Tokyo University of Science, Suwa, Japan
Sten Hanke, IPN Instituto Pedro Nunes, Portugal | AIT Austrian Institute of Technology GmbH, Austria
Michael Bremseth, Winthrop University, USA
Jean-Pierre Cordeau, McGill University Health Centre, Canada
Ivan Evgeniev, TU Sofia, Bulgaria
Matthieu-P. Schapranow, Hasso Plattner Institute, Germany
Abdel-Badeeh M. Salem, Ain Shams University Abbasia- Cairo Egypt

GLOBAL HEALTH Technology Advisory Committee

Marike Hettinga, Windesheim University of Applied Sciences, Netherlands
Shirley Coyle, Centre for Sensor Web Technologies / Dublin City University, Ireland
Anna Cysewska-Sobusiak, Poznan University of Technology, Poland
Kevin Daimi, University of Detroit Mercy, USA
Ravi Janardan, University of Minnesota-Twin Cities, USA
Jose M. Juarez, University of Murcia, Spain
Jochen Meyer, OFFIS Institute for Computer Science, Germany
Vesselin Gueorguiev, TU Sofia, Bulgaria
Miki Sirola, Aalto University, Finland
Daniel Steffen, German Research Center for Artificial Intelligence (DFKI) GmbH - Kaiserslautern, Germany
Maria Taboada Iglesias, University of Santiago de Compostela, Spain
Jari Veijalainen, University of Jyvaskyla, Finland
Pierangelo Veltri, University Magna Graecia of Catanzaro, Italy
Brijesh Verma, Central Queensland University, Australia
Liangjiang Wang, Clemson University - Greenwood, USA
Hans C. Ossebaard, National Institute for Public Health and the Environment - Bilthoven, The Netherlands
Ramesh Krishnamurthy, Health Systems and Innovation Cluster, World Health Organization - Geneva, Switzerland

GLOBAL HEALTH Public/Government/Research Advisory Committee

Christopher Mayer, AIT Austrian Institute of Technology GmbH, Austria
João Quintas, Instituto Pedro Nunes, Portugal
Yoshikazu Seki, National Institute of Advanced Industrial Science and Technology (AIST) - Ibaraki, Japan
Hrvoje Belani, Croatian Health Insurance Fund - Zagreb, Croatia
Stefan J. Darmoni, Rouen University Hospital, France
Chih-Lin Chi, University of Minnesota-Twin Cities, USA
Siddhartha Jonnalagadda, Northwestern University, USA
Bin Song, Oracle, Inc., USA
Egon L. van den Broek, Utrecht University - Utrecht, The Netherlands
Zhongming Zhao, Vanderbilt University / Medical Center, USA
Hua Xu, Vanderbilt University/School of Medicine, The Netherlands

GLOBAL HEALTH 2016 Technical Program Committee

Charlie Abela, University of Malta - Msida, Malta
Sachiyo Aburatani, Computational Biology Research Center - National Institute of Advanced Industrial Science and Technology, Japan
Giner Alor Hernández, Instituto Tecnológico de Orizaba - Veracruz, México
Thierry Badard, Centre for Research in Geomatics - Laval University, Canada
Caroline Barelle, MAAP/ Motion Analysis And Performance, France / National Technical University of Athens, Greece
Hrvoje Belani, Croatian Health Insurance Fund - Zagreb, Croatia
Veselka Boeva, Technical University of Sofia, Bulgaria
Mounir Bousbia-Salah, Badji Mokhtar University - Annaba, Algeria
Pamela Briggs, School of Life Sciences / Northumbria University, UK
Sabin-Corneliu Buraga, "A. I. Cuza" University, Romania
Rita Casadio, Biocomputing Group - University of Bologna, Italy
Bolin Chen, University of Saskatchewan, Canada
Dechang Chen, Uniformed Services University of the Health Sciences - Bethesda, USA
Po-Hsun Cheng, National Kaohsiung Normal University, Taiwan
Chih-Lin Chi, University of Minnesota-Twin Cities, USA
Jim Cimino, University of Alabama at Birmingham, USA
Meyrick Chow, Tung Wah College, Hong Kong
Diane Cook, Washington State University, USA
Michel Cotsaftis, LACSC/ECE - Paris, France
Shirley Coyle, Centre for Sensor Web Technologies / Dublin City University, Ireland
Anna Cysewska-Sobusiak, Poznan University of Technology, Poland
Kevin Daimi, University of Detroit Mercy, USA
Stefan J. Darmoni, Rouen University Hospital, France
Lucio Tommaso De Paolis, University of Salento, Italy
M. Jamal Deen, McMaster University - Hamilton, Canada
Wolfgang Deiters, Fraunhofer Institute for Software and Systems Engineering ISST, Germany
Steven A. Demurjian, University of Connecticut, USA

Ramesh Krishnamurthy, Health Systems and Innovation Cluster, World Health Organization - Geneva, Switzerland
Antonio Lanatà, University of Pisa, Italy
Lenka Lhotska, Czech Technical University in Prague, Czech Republic
Tao Li, Florida International University, USA
Yaohang Li, Old Dominion University, USA
Marius George Linguraru, Sheikh Zayed Institute for Pediatric Surgical Innovation, USA
Igor Ljubi, University College for Applied Computer Engineering - Zagreb, Croatia
Irvin Hussein López Nava, INAOE, Mexico
Samia Loucif, ALHOSN University, UAE
Emilio Luque, University Autonoma of Barcelona (UAB), Spain
Zhiyuan Luo, Royal, Royal Holloway University of London, UK
Malika Mahoui, IUPUI - Indianapolis, USA
Stefano Mariani, Politecnico di Milano, Italy
Roque Marin Morales, Universidad de Murcia, Spain
Elio Masciari, ICAR-CNR, Italy
Massimiliano Masi, Tiani "Spirit" GmbH, Austria
Christopher Mayer, AIT Austrian Institute of Technology GmbH, Austria
Sepideh Mazrouee, University of California - Los Angeles, USA
Jochen Meyer, OFFIS Institute for Computer Science, Germany
Christo El Morr, York University, Canada
Kazumi Nakamatsu, University of Hyogo, Japan
Hongbo Ni, Northwestern Polytechnical University, China
Brendan O'Flynn, Tyndall National Institute, Ireland
Michael Ochs, The College of New Jersey, USA
Carlos Oberdan Rolim, Federal University of Rio Grande do Sul (UFRGS), Brazil
Chika Oshima, Saga University, Japan
Julien Penders, imec / Holst Centre, The Netherlands
Marek Penhaker, VSB - Technical University of Ostrava, Czech Republic
George Perry, The University of Texas at San Antonio, USA
Cinzia Pizzi, University of Padova, Italy
Miodrag Potkonjak, University of California - Los Angeles, USA
David Riaño, Universitat Rovira i Virgili, Spain
Carsten Röcker, Fraunhofer IOSB-INA, Germany
Alejandro Rodríguez González, Centre for Biotechnology and Plant Genomics - UPM-INIA, Spain
Marcos A. Rodrigues, Sheffield Hallam University, UK
Seyed Ali Rokni, Washington State University, USA
Alan Ruttenberg, School of Dental Medicine - University at Buffalo, USA
Saqib Saeed, Bahria University Islamabad, Pakistan
Maytham Safar, Focus Consultancy, Kuwait
Antonio J. Salazar E., Institute for Systems and Computer Engineering - Biomedical Research Center, Portugal
Abdel-Badeeh M. Salem, Ain Shams University Abbasia- Cairo Egypt
Nickolas S. Sapidis, University of Western Macedonia, Greece

Copyright Information

Computational Fluid Dynamics Model for a Closed Infant Incubator

Aziza Hannouch, Vahe Mitilian,
Mohamad Hajj-Hassan, Hassan M. Khachfe
Department of Biomedical Engineering
Lebanese International University
Beirut, Lebanon
{61010036; 6083099}@students.liu.edu.lb,
{mohamad.hajjhassan; hassan.khachfe}@liu.edu.lb

Charbel Habchi
Department of Mechanical Engineering
Lebanese International University
Current Address:
Notre Dame University – Louaize
Zouk Mosbeh, Lebanon
charbel.habchi@ndu.edu.lb

Abstract— **Incubators are essential tools to provide thermal comfort for neonates. The main heat transfer processes in these systems are convection and thermal radiation. The objective of the present work is to analyze the flow structure and heat transfer processes using computational fluid dynamics (CFD). The computational domain consists of a simplified incubator and neonate where two cases are considered. In the first case, the mattress was assumed adiabatic whereas in the second one the mattress is assumed to have a constant heat flux. Case one showed an excess of heat loss from the neonate by both convection and radiation, thus the neonate may suffer from hyperthermia and cold stress. This was solved by using a heat mattress with appropriate fixed heat flux in order to provide the desired thermal comfort.**

Keywords: CFD; biomedical engineering; Incubators; thermal comfort.

I. INTRODUCTION

Healthy neonates can regulate their body temperature while premature and ill infants face difficulties keeping a controlled body temperature without external assistance. Consequently, they can suffer from cold stress and hypothermia that leads to an increase in the rate of morbidity and mortality [1]. This is why there is a need to maintain an optimal thermal environment for the neonate. This can be achieved by using incubators.

Incubators are enclosed chambers with climate-controlled equipment used to keep an infant worm and safe from germs exposure. This is achieved by circulating warm air over the infant skin. The design of efficient incubators is crucial for the neonate survival. Many methods can be used to design and optimize incubators. Numerical simulations, namely CFD (computational fluid dynamics), is one of the most used tools since it does not necessitate expensive experimental work and it can assess locally the incubator environment such as the temperature and velocity [2]. Moreover, experimental studies exhibit risks on the human life especially on neonatal infants.

CFD has been significantly enhanced over the last decade due to advanced engineering technologies [2, 3]. In recent studies, computational fluid dynamics has been greatly used to analyze the heat transfer between the environment and the human body, such as in incubators, panda warmers, or intensive care units [1]. CFD simulation can be very useful in many situations, as they easily allow parametric study and can be of great assistance in the design of more efficient, safe, and reliable medical equipment.

Continuous research is being conducted for designing neonate incubators. The main objective is to provide a controlled temperature in order to increase the neonate survival and growth. In fact, the temperature inside the incubators is correlated with the flow structure which is in its turn affected by the ventilation system. Thus, understanding the relation between these parameters could be a solution to solve many problems inside incubators.

The flow inside the incubator is complex and it is usually considered turbulent. This necessitates the use of a turbulence model to compute the governing equations. The most used turbulence models for incubators are the SST $k-\omega$ and RNG $k-\epsilon$ models [1]. Ginalski *et al.* [2] used the SST $k-\omega$ to study the heat transfer in an infant incubator with and without an overhead screen. In their study, the human body and incubator geometry were generated using CAD application CATIA. Then the CFD software ANSYS Fluent was used to analyze the heat transfer and flow structure.

Many others used the standard or RNG $k-\epsilon$ turbulence model to study the heat transfer and fluid flow in closed incubators or radiant warmers [4-6]. A fair agreement was observed between the results obtained numerically and those from experimental analysis. For example, Sedin *et al.* [7] preformed both experimental studies and numerical simulations and compared the body core temperature and mean skin temperature obtained from both techniques. The maximum deviation between experimental and numerical results did not exceed 0.8°C for the worst case, i.e. a relative error of about 2.2%.

The objective of the present work is to design a new ventilation strategy in incubators using CFD simulations. Two cases are considered here. In the first case, all the incubator walls and mattress are considered adiabatic to mimic double walled incubators. The second case higher inlet air temperature is considered and the mattress was heated with constant heat flux.

The current paper is organized as follows: Section II is devoted to the computational domain and boundary conditions. Section III presents the numerical model and algorithms. The results are then discussed in Section IV. Finally, the conclusion is presented in Section V.

II. Computational Domain and Boundary Conditions

The computational domain consists of an incubator chamber of $700 \times 500 \times 400$ mm^3 with a simplified neonatal model as shown in Fig. 1. There are two inlets (in cyan and green) and two outlets (red and orange) similarly to the Drager Caleo incubator [2, 8]. The total surface area of the infant skin is around 0.1344 m^2 which is similar to the models used by Kim *et al.* [5] and Fic *et al.* [6]. The inlets and outlets have same cross-sectional area of 0.01 m^2.

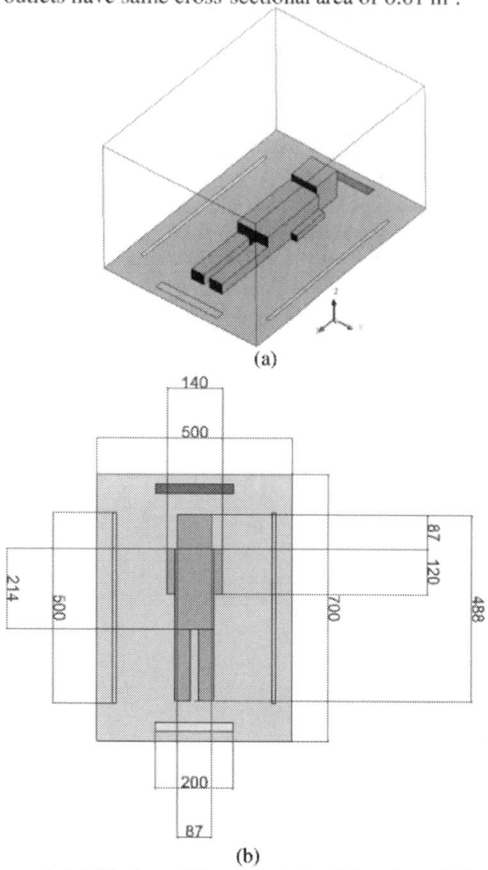

(b)

Figure 1. (a) 3D view of the computational domain and (b) top view with main dimensions

TABLE 1. BOUNDARY CONDITIONS FOR THE TWO CASES

	Case 1	Case 2
Newborn	$T = 36^{\circ}$C	$T = 36^{\circ}$C
Inlet air	$T = 34^{\circ}$C $u = 0.1$ m/s	$T = 34^{\circ}$C $u = 0.1$ m/s
Mattress	Adiabatic	$q'' = 5$ W/m^2
Walls	Adiabatic	Adiabatic

The airflow velocity at the inlets is assumed 0.1 m/s as proposed by Amezzane *et al.* [9] basing on ISO7730 [10]. The pressure at the outlets is considered constant equal to atmospheric pressure. No slip-boundary condition is assumed at all solid boundaries. The incubator walls are assumed adiabatic. Two cases are considered in the present study, as presented in Table 1. For both cases, the baby skin temperature is set to normal temperature of 36°C [11]. The two cases are summarized in TABLE 1. In the first case, the mattress is adiabatic, while in second case, the mattress is heated with a uniform heat flux. A hexahedral mesh is used in the present study consisting of 522,290 cells. The mesh was refined near the wall boundaries to insure y^+ values less than 4 in order to better compute the near wall regions. In the present studies a maximum y^+ of 2.8 was obtained.

III. Numerical Method

ANSYS Fluent [12] is used to compute the Navier-Stokes and energy equations using a second order accuracy. This code is based on cell-centered finite volume discretization. The flow equations are solved sequentially with double precision and a second-order upwind scheme for spatial discretization of the convective terms. The diffusion terms are central differenced and second-order accurate. Pressure-velocity coupling is achieved by the COUPLED algorithm. The $k-\omega$ SST turbulence model is used with a correction for flows characterized by a low Reynolds number. The Boussinesq approximation was adopted to include the buoyancy effects. The radiation process was computed using the Discrete Ordinates (DO) model with 10 iterations. The emissivity of all solid boundaries is assumed equal to 0.9 while the newborn emissivity is set to 0.95 [2]. The residual value 10^{-6} is set as the convergence criterion for the solutions of the flow and energy equations. Beyond this value no significant changes were observed in the velocity and temperature fields and turbulence quantities.

IV. Results and Discussions

The energy rate balance on the neonate body can be written as follows [2]:

$$\Delta q = q_M - q_E - q_C - q_R \qquad (1)$$

where q_M is the metabolic heat generated by the infant, q_E is the heat lost by evaporation, q_C the convective heat loss and q_R the radiative heat loss.

The heat transfer from the newborn to the surrounding environment, especially the radiation and convection rates, is directly related to the temperature and air velocity within the incubator. The convective and radiative heat fluxes are determined from CFD simulations. The first two terms are calculated using the Caleo Drager web application [13]. The baby in the present study has a surface area of 0.134 m^2. The corresponding weight is 1374 g [14]. Thus with this weight and assuming a neonate age of 1 day, this gives $q_M - q_E = 2.24$ W .

For thermal balance the energy difference must be close to zero. In fact, if Δq is negative this means that the baby needs additional heating, and vice versa.

TABLE 2 recapitulates the results for the radiative and convective rates of heat transfer obtained from CFD simulations for case 1 and 2. It is shown than for the case with adiabatic mattress, the heat balance is negative which means that the neonate is losing heat to the environment. This necessitates the addition of an additional heat source to decrease the heat loss.

TABLE 2. ENERGY BALANCE RESULTS FOR CASE 1 AND CASE 2

Heat rate (W)	q_M	q_E	q_C	q_R	Δq
Case 1	3.26	1.02	2.15	5.23	-5.14
Case 2	3.26	1.02	1.10	1.27	-0.13

Fig. 2 to 4 (a) show calculation results for case 1 without heated mattress, while Fig. 2 to 4 (b) show calculation results for case 2 with heated mattress. It can be observed from the streamlines shown in Fig. 2 (a) and (b), that the flow structure for both cases is similar since forced convection is dominant. The flow leaving the inlets is directed upward then the interaction with the upper surface of the incubator redirects the flow downward towards the neonate. Thus a convective heat exchange occurs between the neonate skin and the flowing warm air. Moreover, the temperature of the airflow in case 2 is slightly greater than that in case 1. This explains the reduction in the convective heat loss from the neonate from 2.15 W to 1.1 W. In fact, the average air temperature inside the incubator in case 1 is around 34.8°C while it reaches around 35.3°C in case 2 due to the addition of the heated mattress.

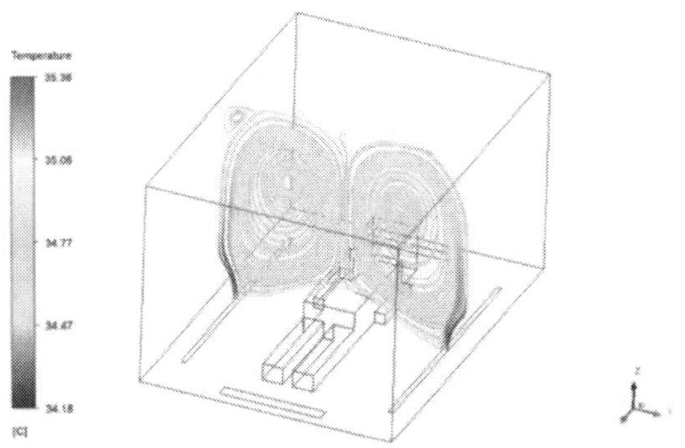

(a)

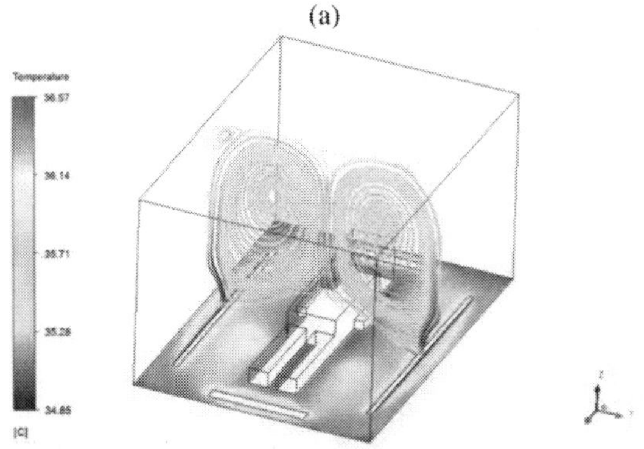

(b)

Figure 2. Streamlines colored by temperature for (a) case 1 and (b) case 2.

The convective heat loss from the neonate for case 1 is shown in Fig. 3 (a) where the values range from -0.31 W/m^2 to 8.12 W/m^2. Positive value means the heat is lost from the neonate while negative value means heat is gained by the neonate skin. Comparing with Fig. 3 (b) for case 2, it is shown that the neonate loses less heat by convection but more heat is added since the air flow temperature is slightly higher in case 2. The convective heat loss is greater at the neonate chest and face due to the impinging flow caused by the two vortices.

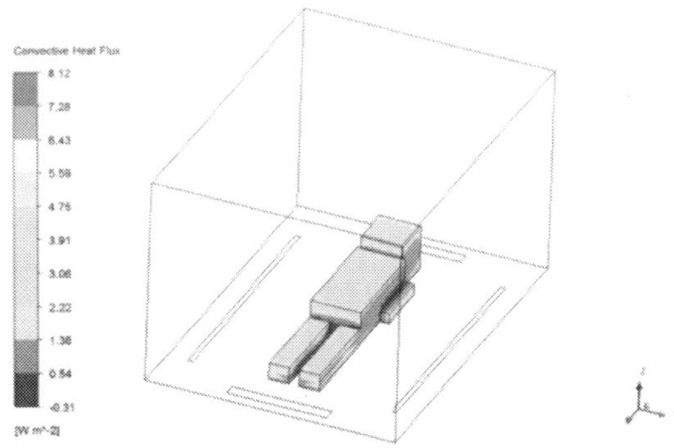

(a)

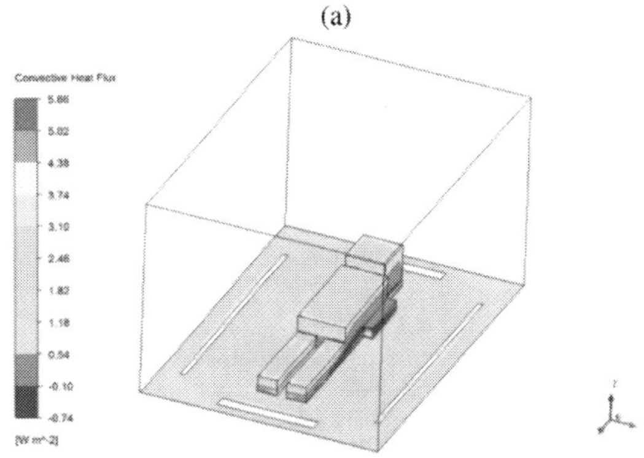

(b)

Figure 3. Convective heat flux on the neonate skin for (a) case 1 and (b) case 2.

The radiative heat flux distribution is presented in Fig. 4, where in case 1 the heat loss by thermal radiation from the neonate is greater than that in case 2. The lowest radiative heat loss is from the inside legs while the maximum is from the baby's face and chest. After the addition of the heated mattress, the radiative heat loss is decreased by almost a factor of 3. Another important issue is that the heated mattress temperature should be almost equal to the neonate skin temperature to avoid conductive heat transfer. This is well obtained here since the heat flux between baby's sides near the mattress tends to zero, avoiding thus overheating issues.

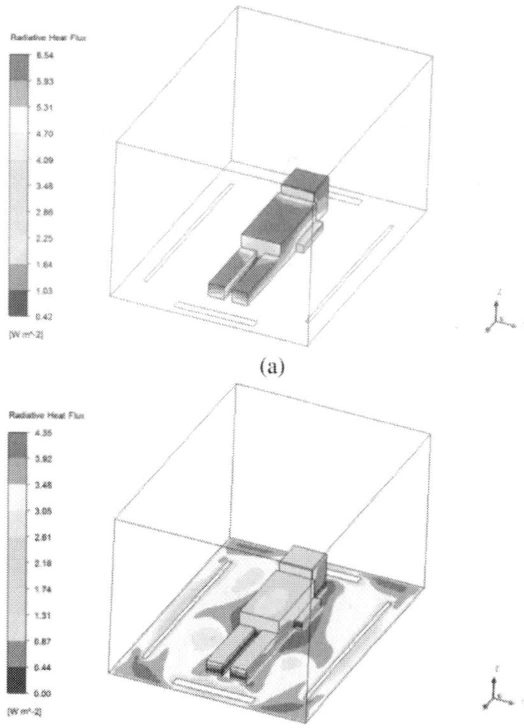

(a)

(b)

Figure 4. Radiative heat flux on the neonate skin for (a) case 1 and (b) case 2.

V. CONCLUSION

Computational fluid dynamics study was performed to analyze the flow structure and heat transfer in a neonate incubator. Two cases where considered: one with an adiabatic mattress and one with a heated mattress. The first case shows that the heat loss from the neonate is high especially for the radiation heat transfer which reaches around 5.23 W. This problem was solved by providing additional heat to the mattress. This could be done for instance by putting an electric resistor underneath the mattress. The heat loss from the neonate with heated mattress where decreased from around 5 W to around 0.1 W. Moreover, the temperature of the mattress did not exceed 36.6°C to avoid burn in the neonate skin.

In future work, the CFD study will account for the humidity comfort and new solutions will be proposed based on the corresponding results.

VI. REFERENCES

[1] L. C. Wrobel, M. K. Ginalski, A. J. Nowak, D. B. Ingham, and A. M. Fic, "An overview of recent applications of computational modelling in neonatology," Philosophical Transaction of the Royal Society of London A, vol. 368, pp. 2817-2834, 2010.

[2] M. K. Ginalski, A. J. Nowak, and L. C. Wrobel, "A combined study of heat and mass transfer in an infant incubator with an overhead screen," Medical Engineering and Physics, vol. 29, pp. 531-541, 2007.

[3] Y. H. Kim, C. H. Kwon, and S. C. Yoo, "Experimental and numerical studies on convective heat transfer in a neonatal incubator," Medical and Biological Engineering and Computing, vol. 39, pp. 114-121, 2001.

[4] M. Rojczyk and I. Szczygiel, "Numerical analysis of radiant warmer," Computer Assisted Methods in Engineering and Science, vol. 20, pp. 237-265, 2013.

[5] Y. H. Kim, C. H. Kwon, and S. C. Yoo, "Experimental and numerical studies on convective heat transfer in a neonatal incubator," Medical and Biological Engineering and Computing, vol. 39, pp. 114-121, 2001.

[6] M. Fic, D. B. Ingham, M. K. Ginalski, A. J. Nowak, and L. Wrobel, "Heat and mass transfer under an infant radiant warmer-development of a numerical model," Medical Engineering and Physics, vol. 32, pp. 497-504, 2010.

[7] G. Sedin, K. Hammarlund, B. Stromberg, "Transepidermal water loss in full-term and pre-term infants," Acta Paediatrica, vol. 31, pp. 305-327, 1983

[8] www.draeger.com (Last accessed on May 2015).

[9] Amezzane, A. Awada, M. Sawan, and F. Bellemare, "Modelling and simulation of an infant's whole body plethysmograph," Medical and Biological Engineering and Computing, vol. 44, pp. 823-828, 2006.

[10] ISO7730, In moderate thermal environments-determination of the PMV and PPD indices and specification of the conditions for thermal comfort, 1993.

[11] S. T. Blackburn, Maternal, fetal and neonatal physiology: a clinical perspective, 4th ed., Elsevier Health Sciences, 2014.

[12] Ansys, "Fluent academic research," 14.5 ed.

[13] Http://legacy.draeger.com/us/heat-balance/index.jsp#help (Last accessed on May 2015).

[14] M. D. Luc brion, M. D. Alan, R. Fleischman, M. D. George, and J. Schwartz, "Evaluation of four length-weight formulas for estimating body surface area in newborn infants," The Journal of Pediatrics, vol. 107, pp. 801-803, 1985.

Rehabilitation of Attention System for Treatment of Traumatic Brain Injury

Maya A. EL Tawil, Amjad M. EL Hadla, Amin Haj-Ali, Hassan M. Khachfe, Mohamad Hajj-Hassan
Departments of Biomedical and Computer Engineering
Lebanese International University, LIU
Beirut, Lebanon
{10930350; 30920005}@students.liu.edu.lb, {amin.hajali; hassan.khachfe; mohamad.hajjhassan}@liu.edu.lb

Abstract— **The main objective of the attention rehabilitation is to increase and enhance the attention level for patients who suffered from traumatic brain injury (TBI) due to certain accidents (e.g., car crash, fall, etc…). Several rehabilitation methods, which perform treatment procedures to patients, have been reported. However, each method has its unique approach to achieve the same goal and to improve the overall result of the patient rehabilitation process. In this paper, the development of an attention rehabilitation system (device and method) are reported. It is based on practicing a video game multiple times to activate the attention area in the patient's brain and to stimulate the patient's willingness to increase his/her effort to win the game. The device and the corresponding game techniques are non-invasive, safe, and to a lesser degree can be practiced autonomously. The patient can practice it at home alone or with minimal supervision. In addition, there are no associated side effects, since no pressure will be applied on the patient and no medication is required during the rehabilitation procedure.**

Keywords- Attention deficits, Rehabilitation, Videogame, Motion sensor, Wiimote, Traumatic brain injury, Wireless data collection

I. INTRODUCTION

TBI is a form of brain injury caused by sudden damage to the brain. Recent studies show that traumatic brain injuries contribute to about 30% of all injury deaths, and the leading causes are falls, vehicles crashes, recreation accidents, and violence [1]. Patients suffering from TBI face difficulties in their daily tasks since a part of their brain has its function disrupted. They also suffer from a lack of cognition for attention, concentration, processing and understanding of information, decision-making, communication, problem solving, organization, reasoning, planning, judgment, memory, controlling impulse and desires [2]. The extent of the disruption in the brain depends on the severity of the injury, whether it is mild or severe.

TBI affects different portions of patient's brain. It mainly affects the attention part leading the patient to lose the ability of gaining particular attention and processing different tasks simultaneously. This results in problems such as carrying on long conversations or sitting still for long periods of time [1]. Several treatment methods are used to perform treatment procedures to such patients and some of them can be performed in hospital, specialized clinics and can happen on an outpatient basis [3]. Some examples of attention and concentration rehabilitation are cognitive rehabilitation therapy, attention process training, and Tomatis attention assist method [4]. These rehabilitation methods are based on training the patients on functional and educational activities by using tools such as classical music or simple exercises. On the other hand, many patients who suffer from TBI will have psychosocial symptoms along with their poor concentration, such as anxiety, anger outbursts, and personality changes [5]. Rehabilitation procedure is going to be hard to perform on such patients, due to the difficulty to deal with their reactions and personalities, especially when the therapy is outside their home and in an unpleasant place such as the clinic or the hospital. Moreover, such patients may refuse to follow directives on what they should do for treatment such as the instruction of the therapist or nurse. In this work, we report the development of a device for the rehabilitation of attention and concentration deficits. The patient can practice at home, and this new method is totally safe, effective and easy.

In Section 2 of this paper, we describe an attention and rehabilitation procedure that is usually done on the TBI patients. In Section 3, a definition and explanation of the software used in this system is stated. In Section 4, the system methodology and components are clearly identified where the proposed workflow is divided into three main parts to simplify the procedure of this system; also, the steps of how the patient can use this device are stated. In Section 5, the results of testing this system on a TBI patient and on a normal person are explained with a brief comparison between these two test cases and determination of the acceptance criteria. In Section 6 we present a summary and conclusion of this new rehabilitation method and state what differentiates it from the new technologies.

II. REHABILITATION OF ATTENTION

Bennett et. al. present the definition of attention as "the ability to focus on certain aspects of the environment that one considers important or interesting and to flexibly manipulate this information" [5]. In other terms, attention is the prerequisite of enhancing the memory and increasing the brain functional activity. It actually plays an important role in the behavior of certain person by making him/her more socially active in communicating with others. In addition, it increases the person's ability to selectively concentrate on specific information from the surrounding while ignoring other things. Furthermore, it induces quick interpretation and understanding of a problem with finding solution faster than others.

The rehabilitation of attention is the ability to restore concentration skills and activate focusing in the brain. Games, activities, and tasks can play an important role in activating the attention, enhancing the focus and the perceptual skills. There are several methods for rehabilitation of attention. However, a general procedure outlined in Figure 1 can be considered typical. This procedure includes six steps that starts with an examination of the patient's brain to determine the amount of the damage. This assessment is based on studying the function of the nervous system and evaluation of cognitive functions such as attention or memory. In other terms, it is called neuropsychological assessment [7] and it must be done prior to the treatment. Based on the results of this test, a series of cognitive and rehabilitation processes are administered to the patient. At the end of this procedure, an evaluation is performed to check the progress and consequences on the patient's brain functions.

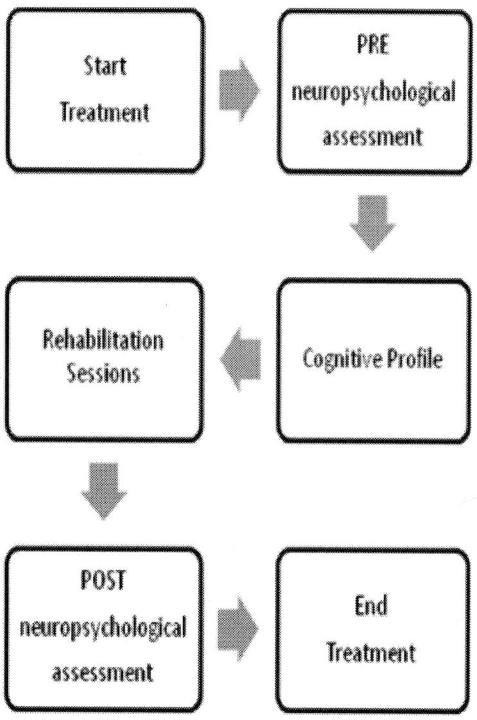

Figure 1. Typical rehabilitation general process [7].

Another important factor to be considered along with the neuropsychological assessment results is the age of the patient. The importance of this factor lies in its direct relation to the patient's brain function. The damage of a young adult brain might be totally different than that of an older person. After collecting all these information from patients, rehabilitation sessions are administered based on their cases. A recording of the patients' cognitive improvement is compiled to produce a comparison between the pre and post neuropsychological assessment.

III. Tools and Methods

This system described in this paper provides a way of introducing patients with lack of cognition resulting from serious brain injury, brain tumor or head trauma to the first steps of rehabilitation. The system includes a hardware control device based on the MindFlex™ headset to record the brain's activities. In addition, a Wii-Mote™ is used as an free guiding device that the patient uses to control the game. The Wii-Mote™ can provide haptic feedback to the patient signaling certain events. The patient uses the headset control along with guiding device in order to move the grasping mechanism in the game (software part of the system) towards a ball with a specific color. By focusing concentration, the patient can initiate the grasping mechanism. The concentration threshold level is usually set by the rehabilitation technician. The aim of the game is to put a certain number of colored balls in their color associated basket. As the patient brain's abilities improve throughout rehabilitation, the technician increases the difficulty level of the game by changing the concentration threshold.

A. *System Architecture*

As described above the brain activities are recorded by the EEG sensors built into the MindFlex™ headset. The headset is outfitted with a Wi-Fi module (ESP8266 based ESP-01). The module retrieves the data stream from the headset via its serial port and echoes the data wirelessly via UDP (User Datagram Protocol) datagrams to a computer. The Wii-Mote™, used to record the patient's hand movement, connects wirelessly as well via Bluetooth. The wireless connections free the patient from any wiring harness and thus allow for more ergonomic set-up.

The computer collecting the data runs the main software application. The application collects the data sent from the headset as well as the patient's hand movement data recorded by the Wii-Mote™.

In addition the application controls both the game's engine and GUI (Graphical User Interface). System block diagram is shown in Figure 2.

B. *Software Application*

The software application is written in Processing™ language. Processing™ is an open source programming language built for the electronic arts, it is a flexible software sketchbook and a language for coding [8].

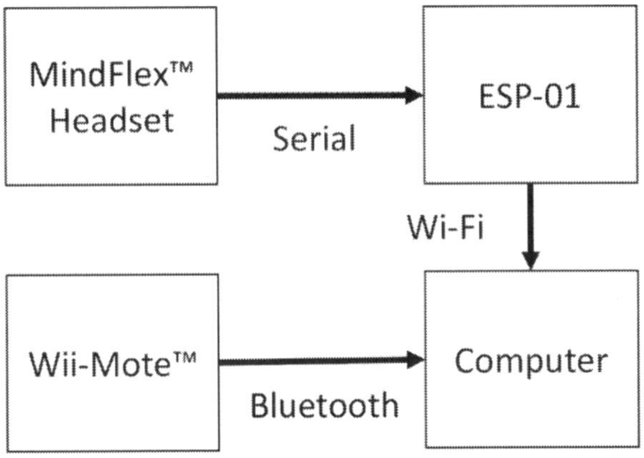

Figure 2. Proposed system architecture

This application can be used at home, hospital, clinics or any comfortable place for the patient. It is important for this application to be used in an area where there is dimmed light interface with the Wii sensor bar in order to have the optimal rehabilitation process and result of the application. The steps to follow are:

1- The patient/therapist will turn on the computer that we have already installed on the game.
2- The patient will wear the MindFLEX headset and turn it on by the button on the left side of this headset.
3- The patient/therapist will open the MindFLEX game icon.
4- The patient/therapist will click on the wireless connection network and click on MindFLEX to connect it with the game software.
5- The Wii sensor bar is connected to the computer through USB port and the Wiimote is paired with the computer by running the Wiimouse program.
6- The patient can sit or stand holding the Wiimote in forward direction toward the Wii sensor bar.
7- The patient can start the game and control the movement of the mouse cursor by holding the button A on the Wiimote, which is equal to the left click on a regular mouse.
8- The patient can press only the button A of the Wiimote in order to move the ball and he/she can move his/her hand left or right to determine the direction that he/she wants to move the ball in.
9- The patient cannot catch the ball unless his/her concentration threshold exceeds the concentration threshold set (by a slider) in the game scheme.
10- The therapist usually determine this concentration threshold based on the patient's progress but any patient can start with setting the concentration threshold to 50,which is the minimum average. The patient can play this game as much as he/she wants per day with no force applied on the patient; in the therapeutic stages, it is better to keep following up the recording of the patient's concentration progress daily.

Depending on the threshold of concentration level taken from the patient, the ball in the game will be captured by the Wii Mote control gesture only if the concentration value above a certain threshold value in various stages of the game. As the patient goes through each stage the stage difficulty goes harder with more obstacles and high concentration level is set while a counter is recording the progress of the patient to provide the data needed for monitoring the effectiveness of the rehabilitation game program on patient activity (see Figure 3).

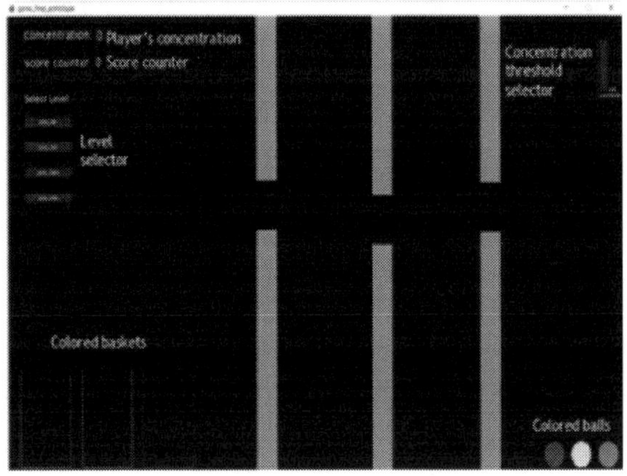

Figure 3. Game graphical user interface

IV. RESULTS

This method was tested on patient who suffers from lack of attention; also it was tested on normal person. As for normal person, he passed the first two levels with less time than that of the patient with attention deficits and the concentration threshold was set to 60, but there were some difficulties to pass the third level. Patient with attention deficits could finish the first stage where no obstacles are set in the game scheme, level 1, and the concentration threshold was set to 50 as it is considered the minimum average that such patients must easily reach in order to catch the ball in the game. It took this patient more time to finish the next level where obstacles are present; however the concentration threshold is still set to 50; and at the end it was so hard for this patient to finish the third level where three obstacles are present in the game scheme. It is important to know that is not necessary to force the patient to finish his/her level, time is not set for the patients and at the end we are trying to encourage him/her to concentrate not to leave him/her disappointed. The purpose of this rehabilitation is to train the patient no matter how much time he/she took in this process in order to get the best result.

V. CONCLUSION

In this work, a new rehabilitation method of attention deficits is devised for TBI patients. It consists of series of brain signals detection, processing and analyzing data. The result of processing showed to be a very effective way to stimulate brain functional activity and enhance the patient's attention level. The use of simple tool to communicate with the game added a specification to this technique and increase its effectiveness compared to other rehabilitation methods. The use of this technique allows the patient to be self-active by coordinating the concentration activity with movement of hand to get through the levels and enhances the divided portion of attention type whereas the other methods lack that coordination and only focus on the focused portion of attention types. The main advantage of this rehabilitation method over other new technological devices like the smart phones with the free applications that may play role in activating the patient's brain, is that they missed the concentration monitor; for example, the puzzle game application, may help the patient to regain his/her concentration skills but there is no recorder for how much he/she is concentrating and if he/she is making a progress or no. When it comes to medical field, a real and concrete values and results must be present in order to diagnose the patient and know how he/she can be treated.

REFERENCES

[1] Privit Blog, Traumatic Brain Injuries: cause and effects, [Online],Available from:http://privit.com/category/blog/2015, retrieved: May, 2016.

[2] D. Neumann and A. Lequerica, "Cognitive Problems after Traumatic Brain Injury", *National Institute on Disability and Rehabilitation Research, U.S Department of Education*, grant no. HI33A060070.

[3] Office of Communications and Public Liaison, "Neurological Diagnostic Tests and Procedures", *NINDS*, NIH Publication No. 05-5380, March, 2005.

[4] W. Mittenberg, R. E. Zielinski, and S. Fichera, "Traumatic Brain Injury: A guide for patients", *Psychotherapy in Private Practice*, pp. 1-13, 1993.

[5] T. Bennett, k. Malia, B. Linton, M. Raymond, and K. Bewick, "Rehabilitation of Attention and Concentration Deficits Following Brain Injury", *The Journal of Cognitive Rehabilitation*, pp. 1-5, March-April, 1998.

[6] Quizlet, [Online], Available from: https://quizlet.com/2016, retrieved: May, 2016.

[7] J. Solana et al. , "Intelligent Therapy Assistant (ITA) for cognitive rehabilitation in patients with acquired brain injury", *BMC Medical Informatics and Decision Making*, DOI: 10.1186/1472-6947-14-58, July, 2014.

[8] B. Fry and C. Reas, Processing, [Online], Available from: https://processing.org/2012, retrieved: May, 2016.

TABLE I. HARDWARE COMPONENTS OF THE SYSTEM

Components	Functionality
 Mind-Flex Headset	Embedded with an EEG sensor measures brain activity and a Neurosky chipset transmit brain signal.
 ESP8266 chipset	Attached to the MindFlex headset. Sends the brain activity signal to the computer via wireless connection.
 Computer	Connect to the ESP8266 chipset to receive brain signal activity. Contain processing software for processing incoming brain data to control designed game.
 Wiimote controller	Connects to the computer through Bluetooth connection. Track the movement of the patient to direct the grapping pointer in the game.
 Wii sensor bar	An IR transmitter to be detected by IR camera within the Wiimote. Used to set the range of motion for the Wiimote controller.

Beyond The Relational Mode: Using Ontology for Medical Data Management

Iyad Zayour
Faculty of Science
Lebanese University
Beirut, Lebanon
email: iyad@alumni.uottawa.ca

Bilal El-Hajj-Diab
School of Computer Science and Software Engineering
University of Wollongong
Wollongong, Australia
emial: b.diab@alumni.uottawa.ca

Bassam Hussein
Department of Industrial Engineering
Lebanese International University
Beirut, Lebanon
email: bassam.hussein@liu.edu.lb

Ali Hage-Diab
Department of Biomedical Engineering
Lebanese International University
Beirut, Lebanon
email: ali.hagediab@liu.edu.lb

Hassan M. Khachfe
Center for Quality Assurance, Institutional Assessment & Scientific Research (QAIASR)
Lebanese International University
Beirut, Lebanon
email: hassan.khachfe@liu.edu.lb

***Abstract*—Software tools that manage structured data are predominantly confined to mainly two tools: spreadsheets and databases. Investigation in the medical field shows that the experience with either tool has not been satisfactory. While a spreadsheet offers flexibility to manipulate the structure of data and can be operated by the end user, it is too limited for extensive data in term of size or complexity. Databases are suitable for moderately complex data that can be large in size, but are costly, take a long time to stabilize and can be only implemented by programmers. These limitations are due to the rigidity of the data model based on arranging data in a tabular form. In order to address this issue, we investigated the frame-based-knowledge-base data model (ontology) from the artificial intelligence literature. This model seems to be more effective and efficient in managing medical research data. To operate using this model, we used Protégé, an open source tool that offered a near database like functionality.**

Keywords-Database; Ontology; Data Management;

I. Introduction

Data management is one of the most classical areas of computing. In the last 30 years, no new data model could challenge the dominance of the relational model. But how good is this model in managing the kind of domains that are conceptually complex, small in size and with continuously changing structure—something we call personal data management? Our investigation in a hospital was that the industrial tools based on the relational model are not satisfactory in managing the kind of data that needs to be collected. Existing database systems have several critical shortcomings that prevent using them directly to process such data. They are too heavy-weight and slow, devoting much complexity to handling tasks that might be irrelevant to the healthcare domain [1].

Interestingly, we identified a more suitable model that has been there for long time in the dust of artificial intelligence (AI) arena. Frame-based-knowledge-base that date back to the early days of AI provided a superior data model for medical data management.

A. *Available Tools and their Data Model*

Practically, available software tools to manage structured data are predominantly confined to two classes: spread sheets and software/database applications.

1) Spreadsheets: On one hand, a spreadsheet is very flexible for creating data models and thus managing data; it is personal built with specific and exact data in mind. Any user with computer literacy can create a tabular list of attributes as columns and actual data as rows. Although a very primitive model, the enormous usability functions and features done by Microsoft on Excel have considerably stretched the limits of what this model can handle. Features like auto filter, auto repeat, dynamic forms and pivot tables touch on many of the functionality that are typically afforded by special purpose software applications.

Yet, these limits did not stretch far enough; data representing several classes/concepts have to be de-normalized to fit the tabular form of a spreadsheet. This de-normalization will increase almost exponentially the redundancy of data. For example, for a one-to-many relation between a professor and students, the complete data of the professor has to be repeated for every student (what an SQL will transparently do in inner join). The common practice and a fundamental principle of building spreadsheets has been normalization, which is a way to build a data model that has the properties of simplicity, non-redundancy, and minimal maintenance [2]. Violating the principle of normalization may create confusion and is far from well-established norms and standards.

2) Databases: On the other hand, software applications based on relational database are suitable to handle moderately complex data that can be very large in size. Theoretically, with enough programming efforts, any data domain can be modeled in relational form with minimum redundancy. However, like any other software, database solutions are costly, take a long time to stabilize and can be done only An by very specialized people; the programmers. This work can be tedious and needs to be done before users can start focusing their attention on the issues that they should be really working on [3]. The end user has little control over adapting the database to the change in requirements that are inevitable in most real world settings.

Microsoft, that stretched the personal Excel into the professional database side, did also try to stretch Access to the more "personal tool" side. But experiences show that there was little success on both sides. Excel still fails when data get to certain size, and Access applications also begin to need full-fledged programmers when requirements bypass the most common scenarios.

It becomes evident that there is a gap between simple knowledge suitable for spreadsheet and costly database application. Thus, there is a need to provide a personalized tool that can manage the complex data space. In order to bridge the gap, we propose and define a personal data management tool that allows the end user to control the structure of data, to enter data and retrieve it without the need for a programmer.

B. Data models

In our opinion, the problem is not in the usability of the tools but rather in the inherent limitations of the data model used in both of spreadsheet and relational databases. This model, based on arranging data in tabular form where all the knowledge space has to be squeezed in between columns and rows, has limited modeling expressiveness all often creating semantic gap with the domain it is modeling. Usually this gap is alleviated either by redundancy as in the spreadsheet or by complex programming as in the database. In this paper, we investigate the little-known frame-based-knowledge-base data model as an alternative approach to commonly used data models. This is an attempt to have a more consolidated base for scattered data; something that would be a must in the near future as elaborated by Tim Berners-Lee (2005): "It is about the data which currently is in relational databases, XML documents, spreadsheets, and proprietary format data files, and all of which would be useful to have access to as one huge database." [4]

Section II outlines the data experience including spreadsheet and database experience. Section III presents a review of the modeling process adapted in this paper. Section IV introduces the concept of ontology. Section V discusses the experience with Protégé. Section VI concludes the paper and provides recommendations for future work.

II. The Data Experience

The data requirement for medical research domain has been for long recognized to be demanding [5]. Our recent experience in a hospital was a perfect example of hitting the relational model limits. The type of medical data that needs to be collected in a hospital is very versatile. Data need to be collected about different objects each with specific attributes in addition to other common attributes. An example of such data and objects includes neuro-imaging parameters (e.g., regional neuro-anatomical and metabolic variations), neuropsychological testing, clinical features, laboratory measures, and genetic studies. Managing this sort of information has been found to be challenging. Next, we describe how the hospital in question tried to manage this data and what kind of challenges it faced.

A. Spreadsheet experience

To save and retrieve data, spreadsheets were initially used in the hospital that we studied. Experience with spreadsheets was satisfying on a smaller size data space. New attributes can be added instantaneously (new columns), data manipulation (add, edit, delete) were trivial and there were no complex space to worry about (e.g., no table relations), they just added columns and rows. Querying was particularly satisfactory using the auto-filter and pivot tables that answered almost all the queries.

Almost any new data requirement could be met by adding new columns, but this could not continue forever. In one spreadsheet that tracks medical imaging information and activities, we reached the "FV" vertical cells (176 cells). At this scale, the spread sheet became poorly useable and prone to error. Navigation activities, such as finding the value of an attribute, understanding the meaning of a column and identifying valid values became increasingly demanding and challenging tasks.

Furthermore, redundancy became a major concern due to the lack of normalization. To add the data of a new image (about 10 new attributes), all other attributes related to the subject need to be repeated. Moreover, as the same column needs to be used for different classes of images having different attributes, it was confusing to identify which attributes (columns) are relevant to the current images and which are not applicable. This issue is related to the lack of inheritance support.

B. Database experience

After the spreadsheet reached its limit, we moved the data from spreadsheet to custom software based on databases. The database approach did not yield any more satisfactory experience. From a development perspective, the suitability issue of relational database models and normalization for complex domains is well-known (e.g., producing a plethora of tables) [6].

More importantly, from the user perspective, the main concern was due to the logistics of database development. This development requires that changes should be contracted to programmers who are not familiars with the data domain. This necessitated a time consuming knowledge transfer task to these programmers.

In our setting, requirements keep evolving as new ideas and observations are brought in. Continuously, there were attributes that are added or removed and values that were reengineered.

Following a rigorous change management process, the change requests created by the evolution in data requirement were more frequent than what the programmers could handle and the time needed in a typical software change was more than users could tolerate; so much so that they eventually, users regressed to spreadsheets.

III. Reviewing Modeling

As we came to the conclusion that the tabular (spreadsheet) and the relational models (database) are unsuitable for our data management requirements, we began looking for alternative models.

In general, the quality of any modeling technique is related to how much it can faithfully represents the interesting knowledge of the domain it is modeling. The notion of semantic gap [7] has been used to describe this sort of quality. A semantic gap, desired to be minimal, refers to the difference between the way knowledge to be encoded is naturally specified and the syntax of the computer representation used to do the encoding. A minimal semantic gap means that the knowledge of a domain can be "paralleled" in a computer representation without applying additional constrains.

In the database literature, the schema design is called the logical design, which is preceded by a more permissible and flexible form of design called the conceptual design. This latter design exhibits the highest fidelity to the real world domain (minimum semantic gaps). To bridge the impedance mismatch between conceptual and relational (logical) model, some relational constrains need to be applied on the conceptual model to produce the logical model.

The conceptual design corresponds mainly to OO-UML (Object Oriented Unified Modeling Language) class diagrams (or entity-relationship diagrams) and it is sometimes called domain modeling or analysis modeling [8]. It models a domain faithfully regardless of the technology that will utilize the model (a UML class diagram can be the input for schema as much as for OO software). Such a design is typically extracted from a conceptual data model, which is a representation of organizational data. The purpose of a conceptual data model is to show as many rules about the meaning and interrelationships among data as possible, independent of any database management system or other implementation considerations [2].

A direct data tool that can parallel a conceptual design should satisfy our modeling requirements. What we needed exactly to satisfy our data management requirement is a personalized tool that can faithfully model our complex data space by paralleling its conceptual model (especially inheritance) and still be personal enough to avoid to the need for costly programming cycles. Note that by personal we mean the ability of the user to modify the data structure himself (like it is in spreadsheets) without the need for a programmer help (like in databases).

IV. Frame Based Knowledge Base (Ontology)

It was interesting to find that one of the oldest technologies that appeared in the early days of computing as part of the artificial intelligence literature was the most satisfying technology. The frame-based-knowledge-base, also called ontology, offers a very semantically rich data and knowledge representation by acting as the semantic mediator for heterogeneous databases [9].

Using one of the many available ontology editors, ontology could represent knowledge and data in a way that accurately parallel the conceptual model. In fact, an ontology editor with sufficient graphic capabilities can replace any conceptual design tool.

Like a class diagram, ontology permits the user to define classes (called concepts in the ontology jargon) and attributes (called slots), even methods in a class diagram can have "daemon" counterparts. An inheritance hierarchy can be created in between classes by means of a binary relation over entities called ISA relation, and even attributes can be inherited to create sub attributes [10].

With an ontology editor, the user defines an ontology for the data space being modeled. However, unlike other forms of design, the implementation of the design is instantaneously available. By allowing the creation of instances of classes, an ontology editor allows the user to enter the actual data in a simple interactive manner. In some ontology editor like Protégé [11] that we choose to use after evaluating several other editors, the UI input forms are dynamically generated by the editor to represents a controlled interface for the data corresponding to the class attributes.

For example, to create a data structure to manage information about a person, we can use Protégé to create a "person" class. Then we can add attributes to concept person such as name, gender, and married status. In return, Protégé will automatically generates a UI form that permits the user to start creating person instances and that contain UI controls to enter data for the person's attributes: a text box for the name, a combo to select "male" or "female" (these are the possible values that were assigned during gender attribute definition) and a check box for the Boolean married attribute.

Creating ontology, thus, was an interwoven and iterative activity of creating conceptual models and entering the data instances for the concepts. The end product is the ontology; an artifact that describes both the abstract conceptual knowledge and the actual data instances. This conceptual annotation and organization of data around the conceptual model has much facilitated the manual navigation and exploration of data. Data instances are explicitly connected to their actual level of abstractions - the concepts (classes). This provides logical data independence, using high-level abstractions to shield the user from the complexity of the underlying hardware and software platforms [1].

V. Experience With Protégé

Although Protégé was satisfactory as a technology platform, we note that several issues arise when users that are used to simple data model have to work with something as complicated as an ontology. Next, we highlight and discuss the advantages and disadvantages of such approach.

A. *On The Intellectual Experience in Ontology Editing*

Conceptual models were never trivial to do, many design decisions are conceptually perplexing such as grouping the decomposition into classes and the location of attributes; unlike the typical simple columns adding in spreadsheets. Not any Excel user would be an ontology editor, as the sophistication of the ontology model requires equal sophistication in users. Classically, this task has been assigned to a specific kind of users commonly called the "knowledge engineer".

Yet, despite the sophistication required in users that reduces the level of personalization (as many user may requires professional helps or training compared to Excel), using Protégé is still very advantageous over the database programming model.

First, any modification of the conceptual model would be of minimum cost. In Protégé, most modifications can be done by simple drag and drop within the integrity constrains (e.g., moving an attribute to super or sub classes).

Second, the utilization of the conceptual design is instant and iterative. The user can be entering new instances just after a class and at least one attribute is created. The cycle between conceptual design and instance creations has been observed to be frequent initially but stabilizes after a while thus reducing wasted time. This iteration is of minimum cost compared to the programming models where any modification will jeopardize the stability of the whole system and incur high costs due to extensive system validation and verification (quality assurance).

B. *Protégé Evaluation*

We found Protégé to be an adequate solution of our current medical data management requirements. Data navigation and exploration has improved drastically. By providing the inheritance capabilities, Protégé solved the issue with the horizontal growth of data in term of attributes that represent the conceptual complexity in data as depicted in Fig. 1. Common attributes are stored in the high level classes and only specialized one are associated with sub-classes.

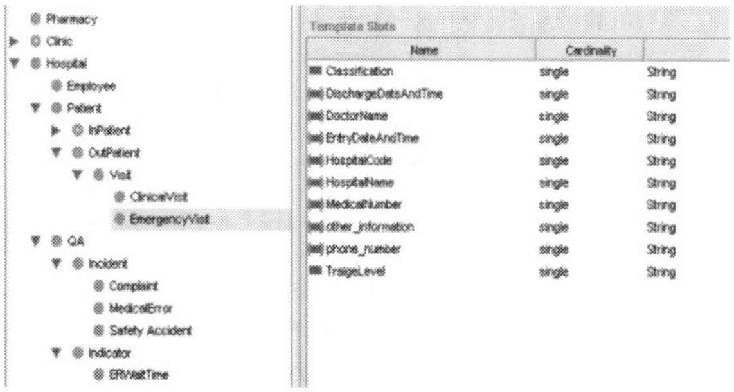

Figure 1. Instance hierarchy and data navigation capability visually created and managed.

Yet, the size of our data is still within the limit of what can be explored and navigated visually. We are not sure how much this approach can handle the vertical growth of data (increase in size) where querying and not navigation would be needed.

In particular, we note that Protégé query mechanism is confined to inter-class conditions (query the instances of one class). This does not represent a shortcoming of the model as much as for the tool. Ontology languages such as RDF, offers full-fledged intra class querying capabilities (compose a condition that utilizes the attribute of more than a class). Although RDF is supported by protégé as a plug-in, the only way to utilize it is by writing queries whose complexity is beyond most end users expectations.

VI. CONCLUSION AND FUTURE WORK

As frame-based-knowledge-base existed for long in the research community, there is an increasing need to transform such a rugged research idea to a tool that meets the everyday requirements and needs of a data user. It is not surprising that frame-based-knowledge-base and ontology mean the same thing since the research community largely failed to envision any utilization of this technology outside the classical artificial intelligence domains and utilizations. A quick look at the knowledge base projects performed with Protégé [12], the leading ontology editor, shows how classical and purely research-oriented established foundations are still dominant for this kind of work.

While excellent data models exist at the basic level of technology, such as XML (Extensible Markup Language) and RDF (Resource Description Framework), which is an XML based ontology language; user-friendly tools for the common tasks in the industry like personal data management are still widely missing.

In our domain (medical research) like many others, data is very valuable and expensive; it is paramount to make the best use of this asset. Rich knowledge representation can help capitalizing on this data by offering models that facilitate finding best conclusions out of this data.

Encouraged by the success we encountered with Protégé in meeting our data requirements, we started to develop a tool that have the advantages of protégé and solve some of its disadvantages. The Personal Knowledge Manager (PKM) is a tool that permits the user to define his data schema and instantly use it to enter data in an Excel like simplicity. It supports inheritance and different table/concept yet it provides a very similar experience to Excel users.

REFERENCES

[1] M. Balazinska et al., "Data management in the worldwide sensor web." IEEE Pervasive Computing, pp. 30-40, 2007.

[2] J. S. Valacich, F. G. Joey, and J. A. Hoffer, Essentials of systems analysis and design. Upper Saddle River, New Jersey, Pearson Education, 2015.

[3] J. Alcala-Fdez et al., "KEEL: a software tool to assess evolutionary algorithms for data mining problems." Soft Computing, pp. 307-318, 2009.

[4] T. Berners-Lee et al., "Tabulator: Exploring and analyzing linked data on the semantic web." In Proceedings of the 3rd

international semantic web user interaction workshop, p. 159, 2006.
[5] H. Andrade and J. H. Saltz, "Towards a Knowledge Base Management System (KBMS): An Ontology-Aware Database Management System (DBMS)". SBBD 1999: 27-39
[6] K. Stoffel et al., A graphical tool for ad hoc query generation. In AMIA'98. American Medical Informatics Association, p. 503, 1998.
[7] D. Merritt, "Best Practices for Rule-Based Application Development Application Development", Microsoft Architect Journal, January 2004.
[8] E. Malinowski and E. Zimnyi "Hierarchies in a multidimensional model: from conceptual modeling to logical representation", Data & Knowledge Engineering, pp. 348–377, 2006.
[9] H. Chen et al., "Towards a semantic web of relational databases: a practical semantic toolkit and an in-use case from traditional Chinese medicine." In The Semantic Web-ISWC, pp. 750-763. Springer Berlin Heidelberg, 2006.
[10] T. Catarci and M. Lenzerini, "Representing and using interschema knowledge in cooperative information systems." International Journal of Intelligent and Cooperative Information Systems, pp. 375-398, 1993.
[11] Stanford University Website. Protégé User Guide. Available online at http://protege.stanford.edu/. Last accessed on June 16th, 2016.
[12] CIM Engineering Website. Highly Effective Distributed Collaboration. Available online at www.cim3.net. Last accessed on June 16th, 2016.

Modeling of Work-Based Access Control for Cooperative Healthcare Systems with XACML

Mohamed Abomhara and Henrik Nergaard

Department of Information and Communication Technology
University of Agder, Grimsatd, Norway
Email: {mohamed.abomhara, henrik.nergaard}@uia.no

Abstract—**Access control is ideal for managing access to information and controlling legitimate user activities by mediating every user attempt to access a system resource. However, in a collaborative environment, the biggest challenges with deploying access control are deciding on the extent and limit of resource sharing as well as difficulty with editing, managing and updating access control policies. In our previous work in the area of access control, we proposed a work-based access control (WBAC) model that strikes a balance between collaboration and safeguarding sensitive patient information. The current study extends on that work by demonstrating how eXtensible Access Control Markup Language (XACML) is used to express WBAC policies. We explain the WBAC model for cooperative healthcare systems, implement the WBAC profile using XACML 2.0, specify permissions and define all authorization policies. We examine the access policies and show how the WBAC model simplifies decentralized administrative tasks (e.g., changing of team members and shifting responsibilities), thus enhancing the practicability of access control in dynamic collaboration environments.**

Keywords–Access control; Collaboration environments; Healthcare.

I. Introduction

Access control policies play an important role in ensuring that the information flow between authorized entities is controlled while preserving resource security in the face of inappropriate access [1]. Access control policies specify which authorized entity (e.g., user or organization) can perform what operations (e.g., read and write) on specific resources (e.g., files on electronic health records (EHRs)).

In collaborative environments such as healthcare, it is not easy for classical access control models like Role-Based Access Control (RBAC) [2] and Attribute-Based Access Control (ABCA) [3] alone to specify authorization constraints due to the complexity of a continuously growing as well as changing number of users and medical records. In addition to a lack of granularity, manageability and flexibility for the specification and maintenance of policies [4], [5]. Moreover, inconsistencies between the access control policies of various individuals or organizations are a common challenge [6]. It is important to understand whether and under what circumstances resources can be shared during collaboration and how collaboration can be achieved securely in the presence of inconsistencies between collaborating participants [7]. Changing participants is another challenge in access control [8]. Access control policies for centralized environments do not address the dynamic changes of participating groups in distributed environments.

The possibility of information leaks caused by improperly designed authorization policies consequently increases. Thus, some extra authorization constraints should be added to traditional authorization mechanisms to prevent information leaks caused by inadequately designed policies. Moreover, access control policies must be flexible and configured to control the dynamic interactions during collaboration [9], [10].

In this study, we demonstrate and implement an access control model for a collaborative healthcare environment to support diverse domains of data authorization management with various constraints. The implantation is built based on eXtensible Access Control Markup Language (XACML) [11] using a Work-Based Access Control (WBAC) model [12], [13], [14] (works by one of the current authors). The aim is to simplify decentralized administrative tasks and thus enhance the practicability of access control in dynamic collaboration environments. WBAC introduces the team role concept, and modified the user-role assignment model from previous works [15], [16]. WBAC handles access control based on collaborative work and team member assignment. The team is segregated into *strategic*, *action* and *management* groups depending on the contributions to the collaborative work.

The remaining parts of this study are structured as follows: Section II presents a background of XACML and usage scenarios of collaboration and healthcare data sharing. Section III demonstrates the modeling structures, authorization constraints, request model and policy model. Section IV presents the experiments and results. Related works, discussion, conclusions and future work recommendations are provided in Section V.

II. Background

In this Section, relevant work underlying the current study is discussed. First, the XACML framework is briefly introduced, followed by concise usage scenarios to better understand collaboration in the healthcare domain.

A. An Overview of XACML

XACML is a standardized policy language by OASIS [11]. It defines the architecture, policies and messages of an access control system. XACML is a powerful and flexible policy language for heterogeneous distributed systems and is a general-purpose access control policy language [17], [18]. According to the reference XACML architecture shown in Fig. 1, the XACML model contains the following main entities [19]:

- *The Policy Enforcement Point (PEP)* is an entity that intercepts a user's request to access a resource. The

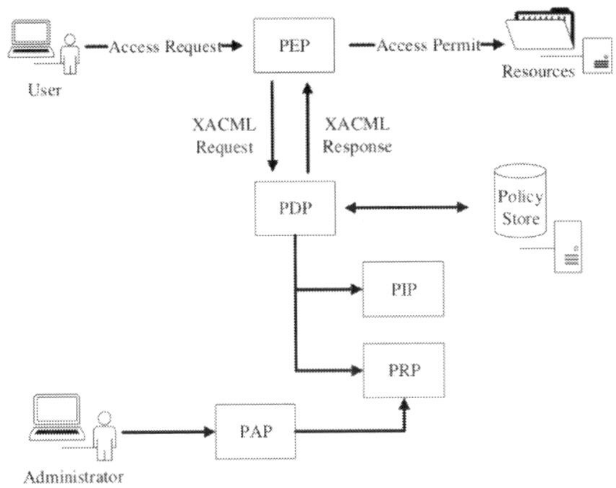

Figure 1. XACML framework

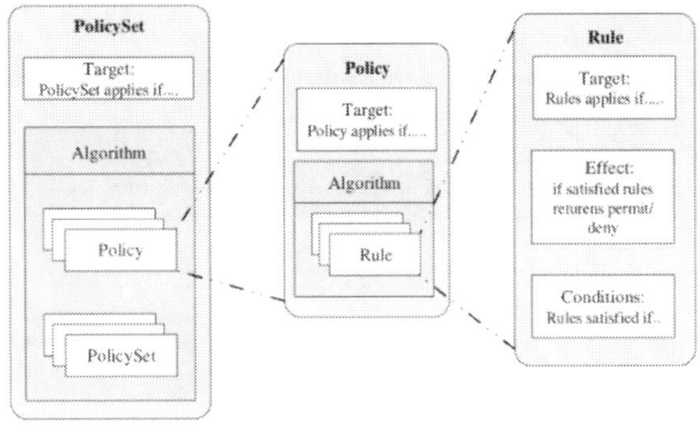

Figure 2. XACML Policy Structure

PEP forwards the request to the PDP to obtain the access decision (permit or deny). PEP then acts on the received decision.

- *The Policy Decision Point (PDP)* is used to evaluate access requests against authorization policies and makes decisions according to the information contained in the request before issuing access decisions.
- *The Policy Information Point (PIP)* acts as the source of attribute values, or the data required for policy evaluation.
- *The Policy Retrieval Point (PRP)* is an entity that stores the XACML access authorization policies.
- *The Policy Administration Point (PAP)* manages the access authorization policies.

The XACML core policy structure (Fig. 2) consists of three components: the rule, policy and policy set [19]. The rule is a fundamental component of an XACML policy. The rule, policy and policy set have a target that PDP uses to quickly find the sub-policy parts applicable to making a decision regarding an access request. The target contains a set of attribute value pairs for matching the subject, resource, action and environment, to check if the given rule, policy and policy set are applicable to a specific request. Several rules are grouped and encapsulated into policies and policies are grouped into policy sets. A rule consists of a condition and an effect that can be either a permission or denial associated with the successful evaluation of the rule. A condition represents an expression that refines the applicability of the rule beyond the predicates implied by its target. The correct evaluation of a condition returns the effect of the rule, while incorrect evaluation results in an error (*Indeterminate*) or the discovery that the condition does not apply to the request (*Not Applicable*).

PDP can use different rules, policies and policy sets to make a decision for a specific request. Therefore, conflict might occur between multiple policies when policies offer different authorization decisions. Thus, XACML provides a set of combining algorithms for combining rules and policies to solve a decision conflict between multiple policies [19]. The most commonly utilized combining algorithms are as follows:

1) *Deny-overrides*: combines the request evaluation result in such a way that if any rule or a policy evaluated to deny, then the request is denied.
2) *Permit-overrides*: combines decisions such that if any rule or a policy evaluates permission, then the decision is permitted.
3) *First-applicable*: combines decisions in such a way that the final decision is made based on the first rule or policy in the policy file.
4) *Only-one-applicable*: This combining algorithm exists only to combine policy sets and policies. It cannot be used to combine rules. It returns the effect of the unique policy in the policy set that applies to the request; whether Deny or Permit.

Based on the combining algorithm used, PDP computes the authorization decision corresponding to the given access request. PDP evaluation is based on the rule, policy and policy set, for which the PDP returns the authorization decision, *permit*, *deny*, *notApplicable* or *indeterminate*. PDP can also returns to PEP a sequence of actions called *obligation* that should be performed in conjunction with enforcing the authorization decision applied to the access request given.

B. An Example Scenario of Collaboration and Sharing of Healthcare Data

Our implementation is modeled based on a typical user case scenario adopted from [20] and shown in Fig. 3. A patient named *Alice* is recently diagnosed with gastric cancer. Surgical removal of the stomach (gastrectomy) is the only curative treatment. For many patients, chemotherapy and radiation therapy are given after surgery to improve the chances of a cure. *Alice* enters a cancer-treatment center at her chosen hospital (e.g., hospital *A*). *Alice* has a general practitioner (*Dean*) whom she regularly visits. Upon her hospital visit, *Alice* also sees an attending doctor (*Bob*) from the same hospital. *Alice*'s health condition has caused some complications, so her attending doctor would like to seek expert opinions and consult about *Alice*'s treatment with different hospitals (e.g., hospital *B*), including *Alice*'s specific general practitioner who is fully informed about *Alice*'s medical history. Note that the invited practitioners are specialized in different areas, where some are specialists and others are general practitioners.

In such group consultation, it is noted that:

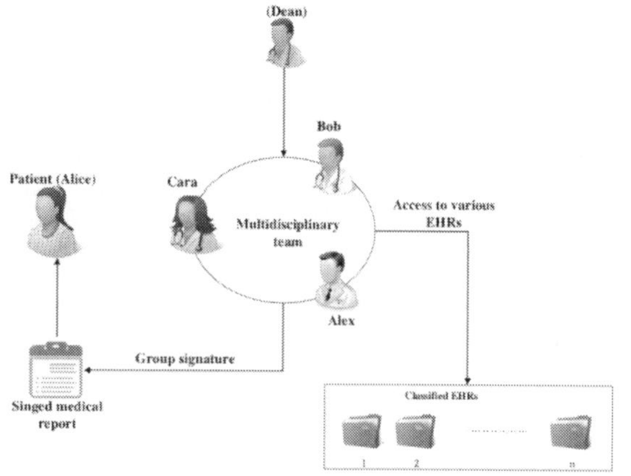

Figure 3. Example scenario of collaboration and sharing of healthcare data

- Several healthcare professionals from different disciplines are involved in various roles to provide patient care.
- The care team are formed dynamically and can be readily changed. For example, when *Alice*'s health condition causes some complications, her attending doctor wishes to seek expert opinions and consult with specialists. As a result of a request for a gastroenterology consultation, we assume a gastroenterologist (*Cara*) will join the care team.
- Every participant needs to obtain some medical records based on the health insurance portability and accountability act (HIPAA) [21] minimal disclosure principle [20], [22].
- Sharing and accessing healthcare records with efficient coordination between healthcare practitioners to perform collaboratively is a critical function in access control models [23]. The main concern regards losing control of sensitive healthcare records while sharing them with multiple parties.

The act of managing the collaborative work in a given scenario must be defined clearly. By default, only the main practitioner (*Dean*) should be aware of the patient's personal information. The three other medical practitioners with supporting roles receive information based on their contributing roles (need-to-know principle). The act of managing a particular collaborative work and how to strike a balance between collaboration and safeguarding sensitive patient information were explained in more detail in our previous works [12], [14], [13].

III. WBAC MODEL IN XACML

The WBAC work model (Fig. 4) [12], [13], [14] postulates that the entire nature of collaboration can be centralized by the *work* concept. Here, each *work* is uniquely identified [13] and is connected to three main components: *personnel*, *patient* and *resources*.

Managing the access control of collaborative work is an interplay between these components. Every resource in WBAC is considered a collaborative entity when it is assigned a

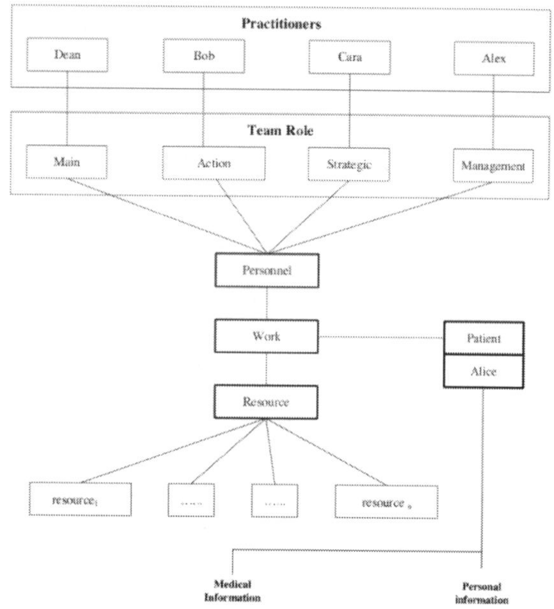

Figure 4. WBAC Work model [14]

workID. The *workID* connects the resource to a corresponding work or project that is done cooperatively. By default, if a resource does not have a *workID*, it implies that it is not a collaborative resource and thus cannot be shared.

Any action that a subject (e.g., healthcare provider) would do on a resource (e.g., patient EHR) is defined entirely within the policy. A dynamic policy with dual inclination is proposed in WBAC [12], [13], whereby the normal policy of enforcing access control is contained within the main policy. On the other hand, any policy that mediates between resource sharing and collaborative work is covered by the collaboration policy. This way, better access control management is achievable. The main policy depends on the roles of the personnel in the organization (e.g., Dean is a general practitioner). PDP only considers the main policy if the personnel possess roles. The collaborative policy is dependent on team roles. In this respect, even if personnel do not have the required roles, they can still gain access upon invitation to collaborate. The team role provides a demarcation between the roles of personnel within a collaboration work and it restricts the role that each team member can have. A person can have various team roles, whereby each is tied to a different collaborative work.

A. Modeling Structures

With the WBAC model, the policy is defined as a tree structure that narrows the combination of attributes presented in a request. Access to a specific resource is granted when the whole policy tree has found possible matches to the request; the result from rule evaluation is then combined upwards to the outer-most policy using the combining algorithm defined at that level. The result is then sent back to the PEP.

The XACML structure of our model is as follows:

1) Subjects, resources and actions are elements defined by identifier/value pairs. Subjects (e.g., healthcare

providers) are entities that send an access request to perform an action (e.g., read or write) on a resource (patient EHRs). The subject is modeled based on the minimal number of attributes required to make different decisions the policy is built to handle. Examples of identifiers are *role*, *employeeID*, *hospitalID* and/or *patientID* (e.g., a patient for whom the physician is responsible), to name a few. For the collaborative part, the information about the subject also includes the team identifier for the current collaboration work. As shown in Fig. 5, physician *Bob* has been assigned the role of attending doctor in the hospital to perform some tasks. He is invited to a collaborative work (*work No 1*) and is assigned the team role *action* to perform some tasks in *Alice*'s treatment.

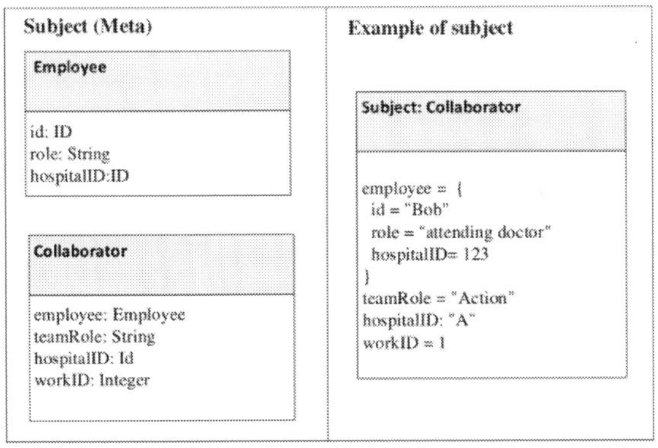

Figure 5. Example of subject attributes

2) *Collaboration members* comprise a group of healthcare providers (specialists or general practitioners) who are invited to a collaborative work (in our case *Alice*'s treatment). Based on the given scenario, *Dean* is responsible for initiating the work and choosing the practitioners (team of doctors) who may be required to attend *Alice*'s consultation and treatment.
Bob, *Cara* and *Alex* joined the team and are assigned team roles based on the required job functions. Table I presents the policy data used as an input for XACML. An action represents the operation that a subject can perform on a resource, e.g., *read* and *write* operations. In our model, we also consider several resource attribute as show in Fig. 6. We also assume the resource are classified into two categories *private* and *protected* (more details about object classification can be found in [13]).

B. Authorization Constraints

We describe the authorization constraints based on the team role classification done in [12], [13], [14] and our usage-scenario (Section II-B) as follows:

- The subject (healthcare provider) who is assigned the primary-doctor role can access both *private* and *protected* resources of the patient for whom he/she is responsible.
- A collaborative work must be active, such that team members can work on it. Assuming the value set

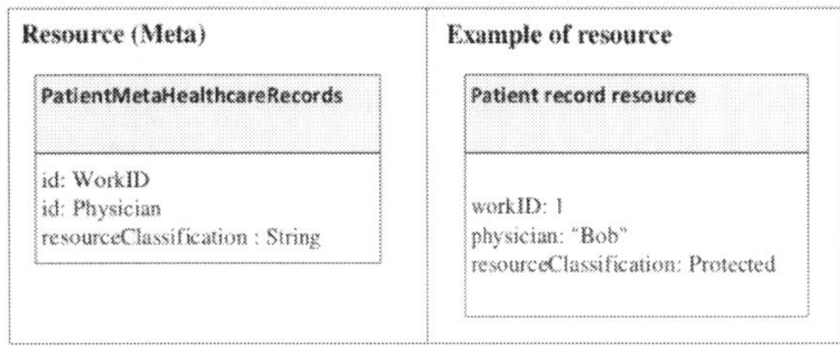

Figure 6. Example of resource attributes

assigned to a work is its identifier, and if there is no work, the field will not be present in a request.

- Only a subject (healthcare providers) who is a member of the care team and is assigned the *action* team role can access *private* and *protected* resources, but only if needed (inevitably). In this model, we assume the healthcare provider who is assigned the *action* team role needs to access private resources because he/she needs to see a patient on a face-to-face basis to perform various tasks related to the patient's recovery. In this respect, there is a need for the healthcare provider to know personal and medical information about the patient to perform his/her duty effectively. Note that in other scenarios, a healthcare provider who is assigned the *action* team role might not need to know private information about the patient.
- Only a subject (healthcare providers) who is a member of the care team and who is assigned the *strategic* team role can access *protected* resources, which are approved for collaboration works. This healthcare provider is predominantly preoccupied with diagnosing the disease, and there is no urgent need for him/her to know the patient's personal information. He/she is responsible for helping the primary doctor to solve the medical case. In fact, he/she is only required to analyze the medical situation and suggest a possible solution. In our model, personnel assigned the *strategic* team role are permitted access only to *protected* resources (e.g., any resources related to the current case of the patient).
- Healthcare providers who are assigned the *management* team role are responsible for coordinating the other team members' interaction by managing meetings and resolving problems with conflicting diagnoses made by other team members. The healthcare provider does not really need to know the patient's personal information. However, he/she must be aware of the patient's medical information to enable coordination. Similar to the *strategic* team role, personnel assigned the *management* team role are permitted access only to *protected* resources. The difference between the *strategic* and *management* team roles is the need for personnel assigned to the *management* team role to have access to team member (healthcare provider) records to be informed of specialist information related to the team members (physicians) in order to coordinate the collaborative work effectively.

TABLE I. Tabular structure of policy data

Subject	Job Function	Team Role	Object Type	Action	Permission
Dean	Primary Doctor	Main role	Private and protected	Read/write	Permit
Bob	General practitioner	Action	Private and protected	Read	Permit
Cara	Gastroenterologist	Strategic	Protected	Read	Permit
Alex	Medical coordinator	Management	Protected	Read	Permit

C. Request Model

The XACML request contains the attributes related to subject, resource and action with their corresponding values. For example, in our case and as depicted in Fig. 7 we have attribute *Subject:Role* and its value *General practitioner*, and attribute *ResourceClassification* and its value *protected* as well as an action value *write*. This information is necessary for authorization decision-making. When PDP evaluates the request against the policy, the attribute names and attribute values are compared according to criteria defined in the policy.

```
<Request>
  <Subject>
    <Attribute AttributeId="subject:id" DataType="string">
      <AttributeValue>Bob</AttributeValue>
    </Attribute>
    <Attribute AttributeId="subject:role" DataType="string">
      <AttributeValue>General practitioner</AttributeValue>
    </Attribute>
    <Attribute AttributeId="subject:collaboration:work" DataType="string">
      <AttributeValue>1</AttributeValue>
    </Attribute>
    <Attribute AttributeId="subject:collaboration:role" DataType="string">
      <AttributeValue>action</AttributeValue>
    </Attribute>
  </Subject>
  <Resource>
    <ResourceContent>
      <record>
        <patient>
          <physician>Dean</physician>
          <work>1</work>
        </patient>
        <classification>protected</classification>
      </record>
    </ResourceContent>
    <Attribute AttributeId="resource-id" DataType="string">
      <AttributeValue>patientRecord</AttributeValue>
    </Attribute>
  </Resource>
  <Action>
    <Attribute AttributeId="action-id" DataType="string">
      <AttributeValue>write</AttributeValue>
    </Attribute>
  </Action>
</Request>
```

Figure 7. Example of an XACML access request

D. Policy Model

The XACML collaboration model begins with a top-level policy set containing one policy for handling a case where the subject is the patient's primary physician and a policy set for the different collaboration cases as shown in Fig. 8.

The top-level policy combines the results based on first applicability, meaning that if the requesting subject is the patient's primary doctor, he/she will get access to records regardless of collaboration. PDP will receive all policies as inputs, where each policy has an element known as "target". The target element's attribute values (subject, resource, action and environment) are matched with the incoming request (Fig. 7) attribute values to decide whether a particular policy is applicable to a given request. If the request attributes match the target's attributes, the policy will be evaluated further. Else, PDP decides the given request is not applicable to the policy.

```
<PolicySet PolicySetId="patient-collaboration" PolicyCombiningAlgId="policy-combining-
algorithm:first-applicable">
  <Target>...</Target>
  <!--
    Policy ensuring that the primary physician has clearance to access medical records
  -->
  <Policy PolicyId="team:manager:doctor:record:access:policy" RuleCombiningAlgId="rule-
  combining-algorithm:permit-overrides">
    <Target>...</Target>
    <Rule RuleId="isPrimaryDoctor" Effect="Permit">...</Rule>
  </Policy>
  <!-- Collaboration Policies -->
  <PolicySet PolicySetId="collaboration:policy:set" PolicyCombiningAlgId="policy-combining-
  algorithm:deny-overrides">
    <Target/>
    <Policy PolicyId="thought:policy" RuleCombiningAlgId="rule-combining-algorithm:permit-
    overrides">...</Policy>
    <Policy PolicyId="actioneer:policy" RuleCombiningAlgId="rule-combining-algorithm:permit-
    overrides">...</Policy>
    <Policy PolicyId="Management:policy" RuleCombiningAlgId="rule-combining-algorithm:permit-
    overrides">...</Policy>
  </PolicySet>
</PolicySet>
```

Figure 8. Screenshot of top-level policy set

Fig. 9 displays a sample policy (part of the defined policy), which ensures the primary physician has clearance to access medical records.

```
<!--
  Policy ensuring that the primary physician has clearance to access medical records
-->
<Policy PolicyId="team:manager:doctor:record:access:policy" RuleCombiningAlgId="rule-
combining-algorithm:permit-overrides">
  <Target>
    <Subjects>
      <Subject>
        <SubjectMatch MatchId="string-equal">
          <AttributeValue DataType="string">doctor</AttributeValue>
          <SubjectAttributeDesignator DataType="string" AttributeId="subject:role"/>
        </SubjectMatch>
      </Subject>
    </Subjects>
  </Target>
```

Figure 9. Example of a policy ensuring that the primary physician has clearance to access medical records

While the target element evaluates the applicability of a policy, the rule element implements the actual authorization logic. The primary physician policy has one rule as demonstrated in Fig. 10, which permits access. If the rule's condition is evaluated as true, the output of the rule will be "permit" where the primary physician field in the resource content patient metadata the same identifier for the subject.

```
<Policy PolicyId="team:manager:doctor:record:access:policy" RuleCombiningAlgId="rule-
combining-algorithm:permit-overrides">
  <Target>...</Target>
  <Rule RuleId="isPrimaryDoctor" Effect="Permit">
    <Target/>
    <Condition>
      <Apply FunctionId="string-equal">
        <Apply FunctionId="string-one-and-only">
          <AttributeSelector
          RequestContextPath="//Resource/ResourceContent/record/patient/physician"
          DataType="string"/>
        </Apply>
        <Apply FunctionId="string-one-and-only">
          <SubjectAttributeDesignator AttributeId="subject:id" DataType="string"/>
        </Apply>
      </Apply>
    </Condition>
  </Rule>
</Policy>
```

Figure 10. Example of a rule that defines the primary physician is permitted access to medical records

Collaboration policies are divided into three sub-policy sets from the main policy set, as shown in Fig. 8. Each policy set is for one specific team role and the rule that applies to this team role. To evaluate collaborative work, the subject *workID* is matched with that of the resource and must be equal for access to be granted and combined with other constraints, such as *read* or *write* effect. An instance of one collaboration policy is shown in Fig. 11. Here, the subject assigned the *strategic* team role is granted access (read access only) to the *protected* resource type if the *workID* matches the active *workID*.

```
<Policy PolicyId="thought:policy" RuleCombiningAlgId="rule-combining-algorithm:permit-
overrides">
  <VariableDefinition VariableId="inSamework">
    <Apply FunctionId="string-equal">
      <Apply FunctionId="string-one-and-only">
        <AttributeSelector
        RequestContextPath="//Resource/ResourceContent/record/patient/collaboration/work"
        DataType="string"/>
      </Apply>
      <Apply FunctionId="string-one-and-only">
        <SubjectAttributeDesignator AttributeId="subject:collaboration:work"
        DataType="string"/>
      </Apply>
    </Apply>
  </VariableDefinition>
  <Target>...</Target>
  <Rule RuleId="protected:resource:rule" Effect="Permit">
    <Target>
      <Resources>
        <Resource>
          <ResourceMatch MatchId="string-equal">
            <AttributeValue DataType="string">protected</AttributeValue>
            <AttributeSelector DataType="string"
            RequestContextPath="//Resource/ResourceContent/record/classification"/>
          </ResourceMatch>
        </Resource>
      </Resources>
      <Actions>
        <Action>
          <ActionMatch MatchId="string-equal">
            <AttributeValue DataType="string">read</AttributeValue>
            <ActionAttributeDesignator DataType="string" AttributeId="action-id"/>
          </ActionMatch>
        </Action>
      </Actions>
    </Target>
    <Condition>...</Condition>
  </Rule>
</Policy>
```

Figure 11. An example of one of the collaboration policy (`strategic` team role policy)

IV. EXPERIMENTS AND RESULT

The WBAC model as described in Section III has been implemented using XACML 2.0. Verifying that this implementation of WBAC can be used as part of an XACML policy was done using the Java SunXACML implementation [24] to run a PDP, testing the policy against different requests. The experiment assumes that the PDP is configured to be deny-biased which means that any response which is *indeterminate* or *not applicable* is seen as a *deny* response.

The WBAC policy was tested by using the attributes based on the data models shown in Fig. 5 and Fig. 6 to build access control requests as shown in Fig. 7. Both valid and invalid values were set for the different attributes to verify that access was permitted and denied correctly.

The experiments showed that the WBAC model granted access correctly to subjects matching the same work as the resource for the expected cases. Invalid request such as a subject *work* with the value 2, while the resource *work* value set to 1. Since the policy is only implemented with rules needed for permitting access when requests is matched the PDP responded with a *indeterminate* answer, which is interpreted as a deny response when the PDP is deny biased.

V. DISCUSSION AND CONCLUSIONS

A. Related Works

Researchers have made the best effort to propose an access control model that balance between security and collaboration requirements [25], [26], [27]. A numerous of research trends on access control approaches have been presented such as RBAC, ABAC, team-based access control (TMAC) [15], task-based access control (TBAC) [28], context-based TMAC (C-TMAC) [16], team task based RBAC (TT-RBAC) [29] and group-based RBAC (GB-RBAC) [30]. In this Section, we compare them to understand better the differences between these approaches. Comparison is imperative and aims at well defining the appropriate access control model for our model. The main evaluation criteria for access control in collaborative system were presented in number of studies [26], [31]. The assessment criteria with respect to healthcare collaborative environments as follows:

1) **Personalized permission**: Patients must be informed of the collaboration and should be given the right to choose who can have access to their records.
2) **Selective confidentiality**: Certain patient information is highly sensitive. Thus, patients should be able to withhold information that remains confidential.
3) **Flexibility and adaptability**: Flexibility is the ability of the access control model to support frequent changes in policy. Whereas, adaptability is used to evaluate the ability of the access control to adapt different healthcare scenarios and environments.
4) **Fine-grained control**: the access control model should support fine-grained subjects, objects and access rights. The granular level at which rules can be applied not only for roles but also for individuals on one or many controlled objects [26].
5) **Groups of users: assignment and revocation**: in collaborative work, common tasks are undertaken by a group of people (team). Therefore, access control model supports the notion of team and allows to specify access rights for teams. Also, the model should has the capability to revoke rights of subjects to access objects.
6) **Policy specifications**: The access control model should allows for scalability and easy extension and modifications of access rights of subjects to access objects.
7) **Policy enforcement**: The access control model should provide means that ensures a correct enforcement of the policy or constraints specification.
8) **Designed for collaborative Healthcare systems**: This criteria show whether or not the access control solutions was designed specifically for collaborative Healthcare systems

Table II summarizes our comparative analysis of the RBAC, ABAC, TMAC, TBAC, C-TMAC, TT-RBAC, GB-RBAC and WBAC models to access control. The table make use of the comparative terminology where "Low", "Medium", and "High" are used to indicate the degree to which the requirement is supported. Also, descriptive terminology such as " Complex" is used to describe the level to which the requirement is supported as well as the standard terminology "Yes", and "No" have been used whenever it is possible to

TABLE II. Access control methods comparison

Access Control models	**Assessment Criteria**							
	1	**2**	**3**	**4**	**5**	**6**	**7**	**8**
RBAC	No	No	Medium	Low	Yes	Yes	Yes	Yes
ABCA	No	Yes	High	High	Complex	Complex	Complex	No
TMAC	Yes	Medium	Medium	Yes	Medium	Yes	Yes	No
TBAC	No	Medium	Low	Low	No	Yes	Low	No
C-TMAC	Yes	No	Medium	Yes	Yes	Yes	Yes	No
TT-RBAC	Yes	No	High	Yes	Yes	Complex	Yes	Yes
GB-RBAC	Yes	No	Medium	Low	Yes	Yes	Yes	No
WBAC	Yes	Yes	High	Medium	Yes	Yes	Yes	Yes

indicate the facilitation or lack of facilitation of the concerned the requirement by the access control model.

WBAC is loosely based on RBAC and ABAC models, and extended with a team role concept. Role and team role are used in conjunction to deal with access control in dynamic collaborative environments. Therefore, the flexibility and adaptability, fine-grained control, policy specifications and policy enforcement are the same as the RBAC. Groups of users: assignment and revocation is similar to TMAC, C-TMAC and TT-RBAC except that in WBAC, the team is classified based on the team role . In WBAC, one team can assigned to a collaborative work in any granularity based on team members' team role. Selective confidentiality is well supported by WBAC because, it is possible to assign a specific object to a each team member in given team based on the object classification and team role. we believe that WBAC handles personalized permission well and meets our expectation of allowing fine-grained access control as well as enhance the practicability and manageability of access control in dynamic collaboration environment. That is, we assume that WBAC is at least as good as RBAC and ABAC, model in this area.

B. Discussion

To prevent any violation of the access control policy of an organization, most classical access control models like RBAC and ABAC define users rights precisely, based on subject and object elements. When several subjects and objects are involved, the subject-object model cannot deliver satisfactory security management. In collaborative environments such as healthcare, it is challenging to predefine all access needs based on the subject-object model. One example of such a situation is explained in our case scenario (Section II-B), which may not be predictable and it would be hard to express the condition of who should join the collaboration and when *Dean* necessitates collaborative support from other parties. Moreover, in deciding on the extent and limit of resource sharing, For instance, in the case of *Alice*'s treatment, which sensitive data should be disclosed to an assisting practitioner so collaboration can be effective, and which should be hidden to safeguard the patient's privacy? Another important matter is the correctness of the policy. Access policy adoption may be limited if the intended policies are not implemented efficiently and consequently thus perform poorly.

WBAC was proposed to address these concerns and support the security and collaboration requirements in access control [25], [26], [27]. The major contributions of the WBAC model include ensuring that access rights are dynamically adapted to the actual needs of healthcare providers and providing fine-grained control of access rights with the least privilege principle, whereby healthcare providers are granted minimal access rights to carry out their duties. In our case scenario, it was noted that general practitioner *Dean* could not solve Alice's case alone. He invited a multidisciplinary team including *Bob*, *Cara* and *Alex* to help. In this team, *Dean* is the core physician in the collaborative work and servers as the group manager. He is responsible for initiating the *work* (*Alice*'s treatment case) and choosing practitioners (group of doctors) who may be required to attend *Alice*'s consultation and treatment. This implies that *Dean* holds the main role. In other words, he owns the initiated collaborative work. Therefore, *Dean* is given a full access (based on his role as primary physician, Fig. 9) with regard to patient-related information. *Bob*, *Cara* and *Alex* are assigned to team roles based on the job function they will perform in *Alice*'s treatment. In our previous work [13], we formally describe and showed how each user joins the team and how each should be assigned at least one team role; a team role can be assigned to none or multiple users in many teams.

In this study, we demonstrated WBAC policies in XACML. We selected XACML because it has been proven to be adaptable to specifying several common access control methods, such as RBAC and ABAC. Our implementation only covers access request for medical records resources, but by using similar matching technique as for the *work* attribute, it is possible to extend this to other polices that are also active during collaboration. An example of this could be for persons with the *management* team role, which should also have access to the personal files like those in the same collaborative work team.

XACML offers extensibility and pluggability which enables the policy presented in this work to be not only a standalone policy, but it could also be a small part of a larger collection of policies. Possible extensions of the base collaboration policy could, for example, be sub-roles of each primary collaboration roles. This could give even more granularity for specific cases for example if a medical employee in the *management* team role.

C. Conclusions and Future Work

A work-based access control (WBAC) model was proposed, which is suitable for collaborative healthcare systems in addressing the subjects of information sharing and security. The aim was to provide a flexible access control model without compromising the granularity of access rights. In this study, we showed how XACML can be used to implement the WBAC model policy and how XACML combining algorithms can be used to manage the inconsistencies between different policy sets. XACML has become very popular in both academia and industry as a standard for combining, maintaining and exchanging access control policies. It is an architecture for evaluating authorization requests and for issuing authorization decisions. The experiments we conducted demonstrated the applicability of XACML to supporting collaborative and distributed domains in sharing access control of specific resources.

In the future, the plan is to formalize access control policies and use automated verification tools to verify interesting properties about the WBAC policy as well as detect consistency using such automated tools. SAT Solver [32] and Alloy [33] are examples of automated verification tools. The plan is also to prototype the functionality to be implemented, to ensure the model's practicality and evaluate the validity of the possible difficulties in managing the model during actual implementation.

REFERENCES

[1] P. Samarati and S. D. C. Di Vimercati, "Access control: Policies, models, and mechanisms," Lecture notes in computer science, 2001, pp. 137–196.

[2] D. F. Ferraiolo, R. Sandhu, S. Gavrila, D. R. Kuhn, and R. Chandramouli, "Proposed nist standard for role-based access control," ACM Transactions on Information and System Security (TISSEC), vol. 4, no. 3, 2001, pp. 224–274.

[3] V. C. Hu, D. Ferraiolo, R. Kuhn, A. Schnitzer, K. Sandlin, R. Miller, and K. Scarfone, "Guide to attribute based access control (abac) definition and considerations," NIST Special Publication, vol. 800, 2014, p. 162.

[4] M. H. Kang, J. S. Park, and J. N. Froscher, "Access control mechanisms for inter-organizational workflow," in Proceedings of the sixth ACM symposium on Access control models and technologies. ACM, 2001, pp. 66–74.

[5] S. Oh and S. Park, "Task–role-based access control model," Information systems, vol. 28, no. 6, 2003, pp. 533–562.

[6] R. A. Shaikh, K. Adi, L. Logrippo, and S. Mankovski, "Inconsistency detection method for access control policies," in Information Assurance and Security (IAS), 2010 Sixth International Conference on. IEEE, 2010, pp. 204–209.

[7] Q. Li, X. Zhang, M. Xu, and J. Wu, "Towards secure dynamic collaborations with group-based rbac model," computers & security, vol. 28, no. 5, 2009, pp. 260–275.

[8] D. Daiqin He and J. Yang, "Authorization control in collaborative healthcare systems," Journal of theoretical and applied electronic commerce research, vol. 4, no. 2, 2009, pp. 88–109.

[9] H. Janicke, A. Cau, F. Siewe, and H. Zedan, "Dynamic access control policies: Specification and verification," The Computer Journal, 2012, p. bxs102.

[10] G. Zhang and M. Parashar, "Context-aware dynamic access control for pervasive applications," in Proceedings of the Communication Networks and Distributed Systems Modeling and Simulation Conference, 2004, pp. 21–30.

[11] T. Moses et al., "Extensible access control markup language (xacml) version 2.0," Oasis Standard, vol. 200502, 2005.

[12] M. Abomhara and G. M. Køien, "Towards an access control model for collaborative healthcare systems," in Proceedings of the 9th International Joint Conference on Biomedical Engineering Systems and Technologies (BIOSTEC 2016), vol. 5, 2016, pp. 213–222.

[13] M. Abomhara, H. Yang, and G. M. Køien, "Access control model for cooperative healthcare environments: Modeling and verification," in IEEE International Conference on Healthcare Informatics 2016 (ICHI 2016) (accepted), 2016.

[14] M. Abomhara and H. Yang, "Attribute-based authenticated access for secure sharing of healthcare records in collaborative environments," in Proceedings of the The Eighth International Conference on eHealth, Telemedicine, and Social Medicine 2016 (eTELEMED), 2016.

[15] R. K. Thomas, "Team-based access control (tmac): a primitive for applying role-based access controls in collaborative environments," in Proceedings of the second ACM workshop on Role-based access control. ACM, 1997, pp. 13–19.

[16] C. K. Georgiadis, I. Mavridis, G. Pangalos, and R. K. Thomas, "Flexible team-based access control using contexts," in Proceedings of the sixth ACM symposium on Access control models and technologies. ACM, 2001, pp. 21–27.

[17] J. F. Alqatawna, E. Rissanen, and B. Sadighi, "Overriding of access control in xacml," in Policies for Distributed Systems and Networks, 2007. POLICY'07. Eighth IEEE International Workshop on. IEEE, 2007, pp. 87–95.

[18] A. X. Liu, F. Chen, J. Hwang, and T. Xie, "Designing fast and scalable xacml policy evaluation engines," Computers, IEEE Transactions on, vol. 60, no. 12, 2011, pp. 1802–1817.

[19] N. Li, Q. Wang, W. Qardaji, E. Bertino, P. Rao, J. Lobo, and D. Lin, "Access control policy combining: theory meets practice," in Proceedings of the 14th ACM symposium on Access control models and technologies. ACM, 2009, pp. 135–144.

[20] R. Zhang and L. Liu, "Security models and requirements for healthcare application clouds," in Cloud Computing (CLOUD), 2010 IEEE 3rd International Conference on. IEEE, 2010, pp. 268–275.

[21] S. J. Dwyer III, A. C. Weaver, and K. K. Hughes, "Health insurance portability and accountability act," Security Issues in the Digital Medical Enterprise, vol. 72, no. 2, 2004, pp. 9–18.

[22] B. Fabian, T. Ermakova, and P. Junghanns, "Collaborative and secure sharing of healthcare data in multi-clouds," Information Systems, vol. 48, 2015, pp. 132–150.

[23] F. T. Alotaiby and J. X. Chen, "A model for team-based access control (tmac 2004)," in Information Technology: Coding and Computing, 2004. Proceedings. ITCC 2004. International Conference on, vol. 1. IEEE, 2004, pp. 450–454.

[24] S. Microsystems, "Official project web site sun's xacml implementation," "[accessed 23-May-2016]. [Online]. Available: http://www.sunxacml.sourceforge.net

[25] B. Alhaqbani and C. Fidge, "Access control requirements for processing electronic health records," in Business Process Management Workshops. Springer, 2008, pp. 371–382.

[26] W. Tolone, G.-J. Ahn, T. Pai, and S.-P. Hong, "Access control in collaborative systems," ACM Computing Surveys (CSUR), vol. 37, no. 1, 2005, pp. 29–41.

[27] H. Shen and P. Dewan, "Access control for collaborative environments," in Proceedings of the 1992 ACM conference on Computer-supported cooperative work. ACM, 1992, pp. 51–58.

[28] R. K. Thomas and R. S. Sandhu, "Task-based authorization controls (tbac): A family of models for active and enterprise-oriented authorization management," in Database Security XI. Springer, 1998, pp. 166–181.

[29] W. Zhou and C. Meinel, "Team and task based rbac access control model," in Network Operations and Management Symposium, 2007. LANOMS 2007. Latin American. IEEE, 2007, pp. 84–94.

[30] Q. Li, M. Xu, and X. Zhang, "Towards a group-based rbac model and decentralized user-role administration," in 2008 The 28th International Conference on Distributed Computing Systems Workshops. IEEE, 2008, pp. 441–446.

[31] S. Alshehri and R. K. Raj, "Secure access control for health information sharing systems," in Healthcare Informatics (ICHI), 2013 IEEE International Conference on. IEEE, 2013, pp. 277–286.

[32] G. Hughes and T. Bultan, "Automated verification of xacml policies using a sat solver," in Proceedings of the Workshop on Web Quality, Verification and Validation (WQVV07), 2007.

[33] X. Wang and A. Rutle, "Model checking healthcare workflows using alloy," Procedia Computer Science, vol. 37, 2014, pp. 481–488.

A Monte Carlo Simulation Based-Approach for Medical Equipment Risks Forecasting

Sahar Ismail, Hassan Nehme, Mohamad Hajj-Hassan
Dept. of Biomedical Engineering
Lebanese International University
Beirut, Lebanon
Email:sahar.ismail93@gmail.com
Email:hassan.nehme@live.fr
Email: mohamad.hajjhassan@liu.edu.lb

Bassam Hussein
Department of Industrial Engineering
Lebanese International University
Beirut, Lebanon
Email: bassam.hussein@liu.edu.lb

Hassan M. Khachfe
Center for Quality Assurance and Scientific Research
Lebanese International University
Beirut, Lebanon
Email: hassan.khachfe@liu.edu.lb

Abstract—**As any technology, medical equipments provide benefits to patients, but they also present significant risks that can affect and compromise patient safety. In healthcare organizations, clinical engineering departments play a vital role in maintaining the safety and reliability of medical equipments. In order to mitigate failures of such equipments and control risks, a proper Medical Equipment Management Program (MEMP) should be established. The purpose of this paper is to forecast risks by using Failure Mode and Effect Analysis (FMEA) method and apply it on Monte Carlo simulation which adds risks analysis to Excel® by @RISK tool. The data of some medical devices is extracted from a hospital's maintenance management system and are identified according to their likelihood, severity, and difficulty of detection. However, the results of this mathematical simulation are integrated in a probability distribution function that enables us to identify medical equipments risks that affect patients, staff, and the work environment and reduce them by providing contingency plans, policies, strategies, and other risk management tactics.**

Keywords-medical equipment; risk management; FMEA; Monte Carlo simulation.

I. INTRODUCTION

As medical technology becomes more complicated, a MEMP must be deployed in healthcare facilities to ensure that medical devices operate according to safety, accuracy, reliability, and performance criteria. Maintenance is one of the most important processes to improve safety, decrease the risk of equipment failure, and minimize the unplanned downtime [1]. However, the money spent on maintenance and failure of equipment is rapidly increasing because of the development of many types of complex medical equipments, the stringent environment they are operating under, and the lack of proper management.

The management program includes a risk management process, which comprises the identification, assessment, and prioritization of risks (defined in ISO 31000 as the effect of uncertainty on objectives) followed by coordinated and economical application of resources to minimize, monitor, and control the probability and/or impact of unfortunate events [2]. The causes of the risks are identified and relevant changes in the system are made accordingly in order to reduce the probability of the error occurring in the future thus reducing harm to patients and providing a safer patient care experience.

Most healthcare organizations follow the manufacturer's recommendations concerning the maintenance program [3]. Campbell and Jardine [4] defined the maintenance excellence as the balance of performance, risk, resource inputs and cost to reach to an optimal solution. In the last decade, maintenance techniques have been notably improving, but most of the healthcare organizations do not profit from the maintenance excellence that Campbell and Jardine established. Moreover, some devices that are similar in their function and design have manufacturer-recommended intervals that vary by one or two factors thus leading to potential financial and time losses. In addition, excessive maintenance can have the same impact as an insufficient level of maintenance; moderation should be the rule.

The status of research on maintenance of medical devices is presented in different models. A model proposed by Fennigkoh and Smith [5] classified equipments according to three parameters: function, physical risks, and maintenance requirements. It was known later as risk-based inclusion criteria that allowed clinical engineering professionals to apply maintenance on limited parts of medical devices.

Ridgway, in the beginning, noted that Preventive Maintenance (PM) is an important factor in terms of reliability, but later on, he indicated that PM does not prevent failure for all equipment and it is not the ideal solution. However, Ridgway provided methods for equipment management such as Reliability Centered Maintenance (RCM). This technique is a corporate-level maintenance strategy that is implemented in any healthcare organization to optimize the maintenance program. Endrenyi [6] indicated that RCM selects the critical component in the equipment and initiates a maintenance management process to correct the failure. Further on, he recognized that RCM is good for indicating the budget and for comparing policies, but it cannot help in achieving real optimization.

According to Hall [7], the two keys of RCM are having a good maintenance history of the medical equipment and the age of the equipment. Furthermore, he indicated that RCM is applicable for younger equipments. However, to balance between preventive and corrective maintenance, Condition Based Maintenance (CBM) is presented to observe and forecast real-time status of machines [7]. CBM is performed when some indicators show that the equipment will fail.

Taghipour et al. [8] presented a multi-criteria decision-making model to prioritize medical devices according to their criticality. Furthermore, in terms of prioritization, Jamshidi et al. [9] developed a fuzzy healthcare failure modes and effect analysis (HFMEA). HFMEA is a systematic method that identifies and prevents equipment problems before they occur by ensuring a safe and clinically desirable outcome [10].

To minimize risk and optimize the cost-effectiveness of medical equipment, a maintenance model is suggested by Khalaf et al. [11]. They evaluated both elements and the results showed poor performance concerning cost and risk management. Therefore, Khalaf et al. [12] developed a new model in order to be used in Palestinian hospitals, which is a mathematical model that uses a mixed integer based approach for maintenance operations schedules for medical equipments. They also proposed a greedy algorithm for an initial solution for the model. In addition, some data extracted from maintenance history of infusion pumps and ventilators were used in a global model that measures the probability of equipment being available and they were analyzed using Matlab. However, this model was validated by developing a model that measures the survival of equipment as function of maintenance and age of equipment using survival analysis approach.

The studies reported above proposed models that share a common theme; different risks are calculated using a single measure that is defined and applied to lead safety, performance inspections, and preventive maintenance activities. These models are simple to use and effective in reducing general risks; yet they lack the ability to identify specific risks. They are far from achieving optimal risk minimization. Also, research into comprehensive frameworks for prioritizing critical medical devices or outsourcing of medical device maintenance is still in its infancy. Researchers should apply new risk-based maintenance models including different new uncertainties to replace the traditional empirical models. Existing advances in risk management for other disciplines should be investigated and taken advantage of.

In our model, a Complete Risk and Decision Analysis Toolkit from Palisade: "The Decision Tools Suite" is used. It is an integrated set of programs for risk analysis and decision-making under uncertainty that runs under the popular Microsoft Excel®. The main tool that was used is @RISK, which adds risk analysis to Excel® using Monte Carlo simulation. The Monte Carlo simulation is a technique used to understand the impact of risk and uncertainty in financial, project management, cost, and other models to identify risks related to medical equipments [13]. FMEA method was also used to prevent failure of equipments. Data related to maintenance and failures of equipments was obtained from a Lebanese hospital to apply it in our proposed model in order to verify and validate its functionality, applicability, and performance.

The proposed methodology is presented in Section II. The implementation process is presented in Section III. This latter, includes collecting data, and integrating FMEA method using Monte Carlo simulation. This is followed by results and discussion summarized in Section IV. Finally, a conclusion and our further expectations are presented in Section V.

II. METHODOLOGY

Medical devices are used in healthcare organizations to support patient care in terms of health and safety. Currently, modern medical devices are complex and operate under severe conditions because of the rapid development and evolution of equipments due to substantial advances in technology. Most of the existing strategies in hospitals have difficulties and challenges in identifying risks and applying optimal risk reduction activities because they lack proper management processes. Therefore, a well-operated management process could be expected to enhance the functioning of medical devices across different healthcare organizations.

The proposed model is meant to identify and assess risks of medical equipments according to q mathematical approach using different parameters. It starts with collecting data concerning medical devices from the healthcare institution in question; a Lebanese hospital in our case. The needed numbers such as the likelihood, detectability, and impact of medical equipment failure are then extracted and analyzed.

There are several methods to calculate the risk value, yet the FMEA method is used as the preferred choice in the current model. FMEA is selected among other methods because it contributes to improved designs for products and processes, to cost savings, and to the development of control plans, testing requirements, optimum maintenance plans, reliability growth analysis and related activities [14]. Furthermore, it is a well-established and widely used method in many other disciplines and has been proven to be effective and efficient by numerous researchers. The FMEA procedure starts with determining the ways in which the input can go wrong, and then determining effects for each failure mode. After that, it identifies potential causes for each mode and list current controls for each cause. Consequently, a risk priority number can be determined and contingency plans and actions should be developed accordingly.

After applying the FMEA method, it will be integrated in Monte Carlo simulation tool that includes @Risk toolkit. @Risk adds risk analysis to Excel® using Monte Carlo simulation. Then the simulation will be performed and the results will be assessed to draw a conclusion.

The methodology may be summarized in the following flowchart:

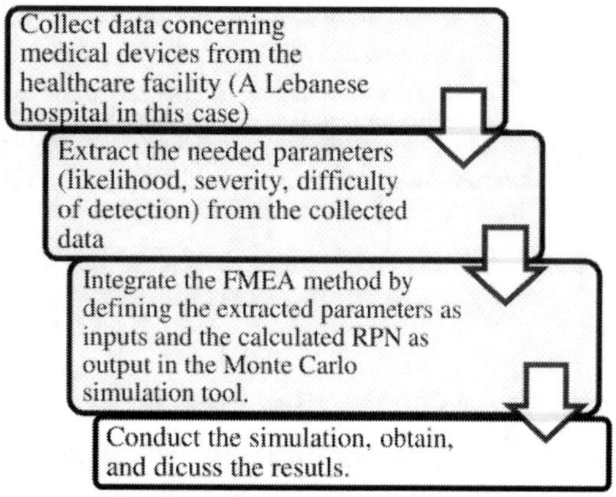

Figure 1. The proposed methodology

Fig. 1 summarizes the required steps to accomplish our evaluation. Such assessment requires some parameters and equations. The derivations of all those relations are explained in the subsequent sections.

III. Implementation Process

A. *Collecting and Extracting Data*

First, to apply the FMEA method, specific data concerning medical devices is collected.

Likelihood of the medical device in this case is the probability of failure of the machine. Fig. 2 shows the number of repeated failures per year with respect to medical devices. These numbers are then converted to a scale of 1-10 as shown in Table I using the following equation:

Number of repeated failures*(10/ Highest number of repeated failures)

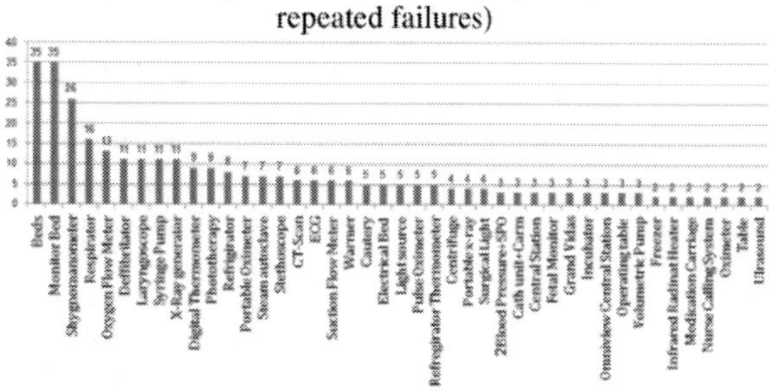

Figure 2. Number of repeated failure.

The scores of likelihood of medical devices failures are assigned according to the following criteria [15]:

{1, 2}: Improbable, manifestations of the hazard are very unlikely

{3, 4}: Remote, manifestations of the hazard are possible but not likely

{5, 6}: Occasional, some manifestations of the hazard are likely to occur

{7, 8}: Probable, hazard will be experienced

{9, 10}: Frequent, hazard likely to occur

Severity of medical device is defined as the extent to which the defect of equipment can affect patients. The scores of severity are assigned according to the following criteria [15]:

{1, 2}: Negligible, no significant risk of injury

{3, 4}: Minor, potential for minor injury

{5, 6}: Moderate, potential for minor injury

{7, 8}: Critical, potential for severe injury

{9, 10}: Catastrophic, likely to result in death

Detection is the ability of the current control scheme to detect and then prevent a hazard from occurring. The scores of detection are assigned according to the following criteria [11]:

{1, 2}: Almost certain (detection probability is between 81-100%), potential hazard will almost certainly be detected

{3, 4}: High (detection probability is between 51-80%), high chance that potential hazard will be detected

{5, 6}: Moderate (detection probability is between 26-50%), moderate chance that potential hazard will be detected

{7, 8}: Low (detection probability is between 10-25%), low chance that potential hazard will be detected

{9, 10}: Remote (detection probability <10%), very remote chance that potential hazard will be detected

TABLE I. EXTRACTED PARAMETERS.

Equipment	Likelihood	Severity	Difficulty of Detection
Beds	10.00	6	1
Sphygmomanometer	7.43	5	1
Defibrillator	3.14	10	4
Ultrasound	0.57	3	3

All values of likelihood, severity, and difficulty of detection of equipment were provided by the hospital. Table I shows the scores of likelihood, severity, and difficulty of detection for twenty-six medical devices on a scale of 1-10.

B. *Integrating FMEA Method in Monte Carlo Simulation Tool*

The parameters extracted from the collected data will be employed in a systematic technique called FMEA.

FMEA is one of the first highly structured, systematic techniques for failure analysis. It was developed by reliability engineers in the late 1940's to study problems that might arise from malfunctions of military systems [16]. It is a step-by-step systematic approach for identifying all possible failures in a design, a manufacturing or assembly process.

Failures are prioritized according to how severe their consequences are, how likely they may occur and how difficult is it to detect them. The main purpose of the FMEA is to take preliminary actions to reduce failures, starting with the highest priority ones [17].

In order to quantify the risk value, a Risk Priority Number (RPN) is used as a measure when assessing risk to help identify critical failure modes. The RPN values range from 1 (absolute best) to 1000 (absolute worst). It is the product of three ratings on a scale of 10 (likelihood of occurrence, severity of impact, and difficulty of detection):

$$RPN = Likelihood\ x\ Severity\ x\ DifficultyofDetection$$

Table II illustrates the extracted parameters and the calculated RPN for each equipment:

TABLE II. CALCULATED RPN.

Equipment	Likelihood	Severity	Difficulty of Detection	RPN
Beds	10.00	6	1	60.00
Sphygmomanometer	7.43	5	1	37.15
Defibrillator	3.14	10	4	125.60
Ultrasound	0.57	3	3	5.13

After calculating the risk priority numbers, the model is now ready to be integrated in the @Risk simulation tool. The first step is to insert Table II in an Excel® sheet and define inputs (likelihood, severity and difficulty of detection) as normal distributions. Usually, high standard deviation is selected in situations where resources are limited or gathering real data would be too expensive or impractical. In this situation, the data is extracted from a real hospital management system, hence a very small standard deviation is selected (0.1); as depicted in Fig. 3:

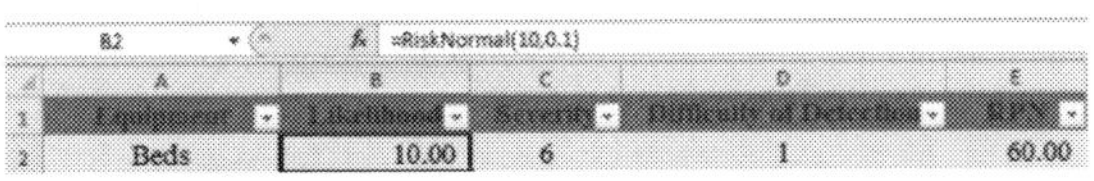

Figure 3. Definition of inputs as normal distributions.

RPN is the output in our model; Fig. 4 illustrates how RPN is defined as an output in the Monte-Carlo simulation tool "RiskOutput("RPN")":

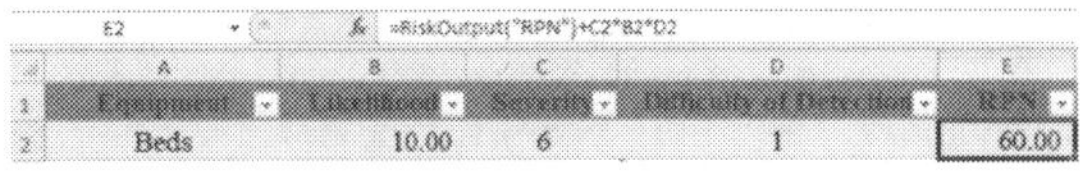

Figure 4. Adding @Risk output.

C. Simulation and Results

@RISK monitors a set of convergence statistics on each output distribution during a simulation. During monitoring, @RISK calculates these statistics for each output at selected intervals (such as: every 1000 iterations) throughout the simulation.

As more iterations run, the amount of change in the statistics becomes less and less until they reach the Convergence Tolerance [18].

Convergence tolerance specifies the tolerance allowed for the statistics being tested. For example, the current applied settings specify that the estimated mean of each output is simulated within 3% of its actual value [18].

In our model in Fig. 5, we will be performing 5000 iterations in one simulation:

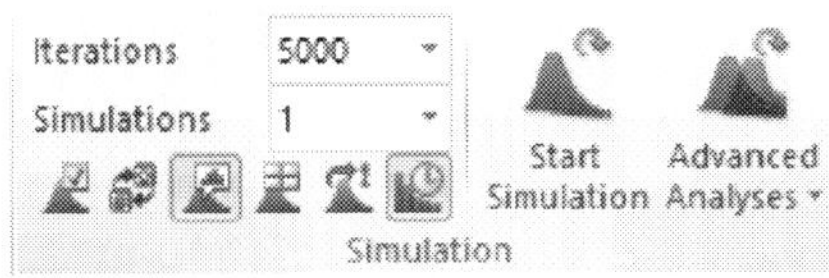

Figure 5. Changing the number of iterations and starting simulation.

At the end of the simulation, the results are integrated in a probability distribution function. A probability distribution is a statistical function that describes all the possible values and likelihoods that a random variable can take within a given range [19]. This range will be between the minimum and maximum statistically possible values, but where the possible value is likely to be plotted on the probability distribution depends on a number of factors, including the distributions mean and standard deviation.

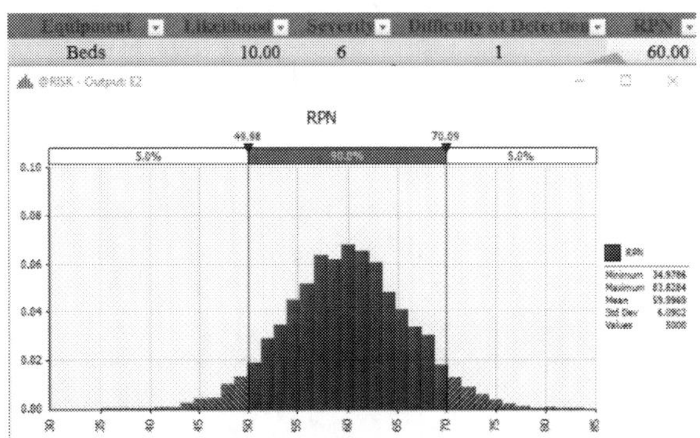

Figure 6. Results after simulation.

Fig. 6 illustrates one example of the results obtained; the risk priority number of hospital Beds (60) is centered between 49.98 and 70.09 for 90% of the probability distribution. The x-axis represents the possible risk priority numbers and the y-axis represents the probability of occurrence for each probable RPN incrementing by 0.02 on a scale of 0.1.

IV. Results and Discussion

The result of the Monte Carlo Simulation via @RISK is a probability distribution. Figs. 6 and 7 show the probability density for the chosen examples: beds and defibrillator.

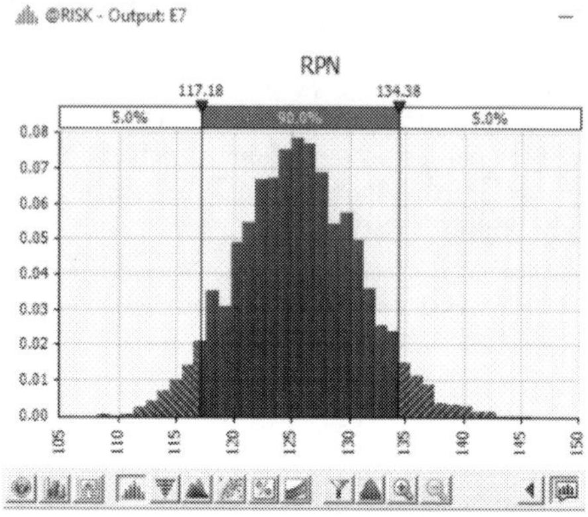

Figure 7. Probability distribution for defibrillator.

TABLE III. SUMMARY OF THE RESULTS.

	Beds	Defibrillator
Minimum	34.97	108.23
Maximum	83.82	145.19
Mean	59.99	125.6

The results presented in Fig. 6, Fig. 7, and Table III are interpretable as follows:

1. The mean figure for RPN will be 60 for beds and125.6 for defibrillator. That means, the simulated result will be equal to the original calculated RPN.
2. The minimum figure for RPN will be 36.57 for beds and 108.23 for defibrillator. This means that the minimum probability will be lower than the calculated RPN by 23.43 for beds and by 17.37 for defibrillator. But these figure are the bottom lines and will only be achieved if all negative circumstances would occur. Hence, with a probability of 5 %, the figure for RPN will fall low to 36.57 and 108.23. In other words, with a probability of 95 % the RPN will not fall below these numbers.
3. The maximum value for RPN will be 85.69 for beds and 145.19 for defibrillator. That is, the maximum probability will be higher than the calculated RPN by 25.69 for beds and by 19.59 for defibrillator. But these figures are the upper limits and will only be achieved if all positive circumstances would occur. Hence, with a probability of 95 %, the figure for RPN will not exceed 85.69 and 145.19. In other words, with a probability of 5% the RPN will exceed these numbers.

An additional evaluation is possible to show where an individual risk has a main influence of the final risk priority number. Figs. 8 and 9 show the results of those evaluations as regression coefficients. This indicates that the difficulty of detection has a huge influence of the RPN of beds and the likelihood has the higher influence on the RPN of the defibrillator. Therefore, these risk factors have to be monitored very carefully within an effective healthcare management system.

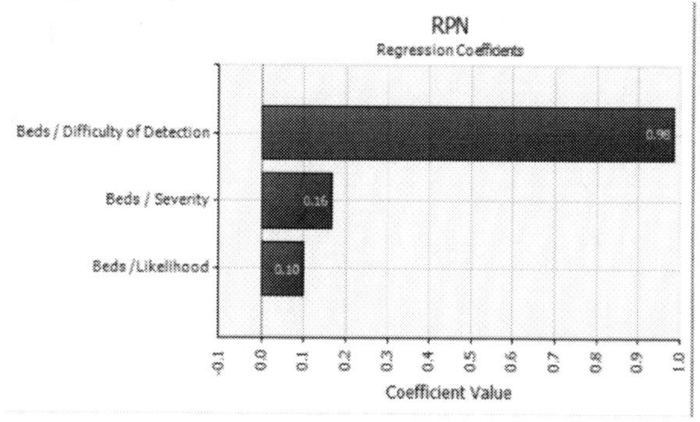

Figure 8. Regression coefficients for beds.

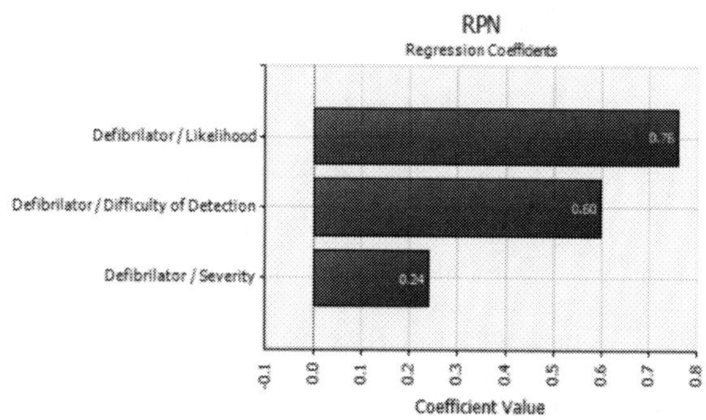

Figure 9. Regression coefficients for defibrillator.

Finally, a risk severity matrix is employed to raise awareness and increase visibility of risks so that proper decisions on certain risks can be made. The risk matrix is shown in Fig. 10. Once the risks have been placed in the cells of the matrix that corresponds to the appropriate likelihood, severity and difficulty of detection, it becomes visibly clear as to which risks must be managed at what priority. This is a well-known and widely used tool in the world of risk management.

Each of the risks will fall under one of the categories, for which different colors have been used. Here are some details on each of the categories:

High: The risks that fall in the cells colored in red are the risks that are most critical and that must be addressed on a high priority basis. Example: 'D' Defibrillator.

Medium: If a risk falls in one of the yellow zone, it is best to take some reasonable steps and develop risk management strategies in time, even though there is no hurry to have such risks dealt with early. Example: 'B' Sphygmomanometer and 'C' Beds.

Low: The risks that fall in the green cells can be minimally monitored and managed as they usually do not pose any significant problem. However, if some reasonable steps can help in fighting these risks, such steps should be taken to improve overall performance Example: 'A' Ultrasound Machine.

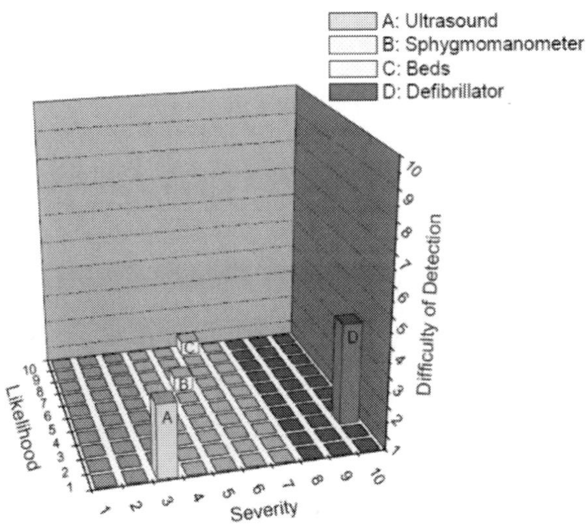

Figure 10. Risk severity matrix.

After analyzing the results, some recommendations could be deduced to reduce risks such as having alternative or redundant devices in the healthcare facility, pay special attention to the to the life span of the equipment and its working hours when purchasing used devices, and to have a well operated maintenance program.

V. CONCLUSION AND FUTURE WORK

The rapid evolution of medical equipments had a huge impact on the improvement and progress of medical services. Accordingly, medical devices are expected to operate under stringent safety, accuracy, and reliability criteria to ensure a protected and efficient environment for patients, staff, and the surrounding work setting. As such, this research work provided a new methodology for identifying and assessing risks based on a mathematical approach and not only empirical ones. This method results in a more precise scheme that would most likely reduce the risks resulting from medical equipments and further provide proper management practices in healthcare organizations. Moreover, this model can be integrated in healthcare facilities to identify and forecast risks according to risk distribution of Monte Carlo simulation and risk severity matrix that classifies and prioritizes medical devices risks.

This proposed assessment maybe further enhanced to achieve risk response development, and risk response control of medical equipment by developing a complete tool that can be used in the medical equipment industry across the world. Thus, manufacturers, organizations, and clinical engineering departments can use this tool in planning for maintenance and for the development of medical equipment. Also, it can be deployed as monitoring system in service at healthcare facilities where it can provide real time data on the risks of operating medical equipment. Another venue for future work in this area would involve further research in the field of optimal outsourcing of medical devices.

REFERENCES

[1] K. Wilson, K. Ison, and S. Tabakov, Medical Equipment Management: CRC Press, 2013.

[2] D. Hubbard, The Failure of Risk Management: Why It's Broken and How to Fix It.: John Wiley & Sons, 2009.

[3] A. Jamshidi, S. Abbasgholizadeh-Rahimi, and D. Ait-Kadi,"Medical Devices Inspection and Maintenance;Literature Review," in Industrial and Systems Engineering Research, Canada, 2014.

[4] J. Campbell and A. Jardine, Maintenance Excellence:Optimizing Equipment Life-Cycle Decisions.: CRC Press, Feb 13, 2001.

[5] L. Fennigkoh and B. Smith, Clinical equipment management.JCAHO PTSM Series, 1989, vol. 2.

[6] J. Endrenyi, "Aging maintenance and reliability," IEEE Power & Energy Magazine, pp. 59-67, May 2006.

[7] A. Hall, Risk-based approach to device management has multiple pathways.: Biomedical Instrumentation & Technology, 2006, vol. 40.

[8] S. Taghipour, D. Banjevic, and A. Jardine, "Prioritization of Medical Equipment for Maintenance Decisions," Journal of Operational Research Society, vol. 9, no. 1666-1687, p. 62, 2010.

[9] A. Jamshidi, S. Abbasgholizadeh-Rahimi, D. Ait-kadi, and A. Ruiz, "A risk-based Maintenance Strategy using Fuzzy HFMEA for prioritization of Critical Medical Equipment.," in Proceedings of the 1st Annual World Conference of the Society for Industrial and Systems Engineering, Washington, USA, 2012.

[10] Maintenance Assistant™. (2015) MA CMMS. [Online].https://www.maintenanceassistant.com/reliability-centered-maintenance/

[11] A.B Khalaf, Maintenance model for minimizing risk and optimizing cost-effectiveness of medical equipment in Palestine.:JClinEng, 2004, vol. 14.

[12] A.B Khalaf, K. Djouani, Y. Hamam, and Y. Alayli, "Evidence-based mathematical maintenance model for medical equipment," in International Conference on Electronic Devices, systems and Applications, 2010.

[13] Risk Analysis and Simulation. New York, USA: Palisade Corporation, September, 2010.

[14] L.S. Lipol and J. Haq, "Risk Analysis Method: FMEA/FMECA in the Organizations," International Journal of Basic & Applied Sciences, vol. 11, October 2011.

[15] lloyd. (2010, August) Medical Device and Diagnostic Industry.[Online].http://www.mddionline.com/article/risk-analysis-beyond-probability-and-severity

[16] U.S. Department of Defense, Procedures for Performing a Failure Mode, Effects and Criticality Analysis., 1949.

[17] ASQ (American Society for Quality). [Online]. http://asq.org/learn-about-quality/process-analysis-tools/overview/fmea.html

[18] Palisade Corporation, Guide to Using @Risk: Risk Analysis and Simulation,add-In for Microsoft® Excel, 2010.

[19]Investopedia.[Online].http://www.investopedia.com/terms/p/probabilitydistribution.asp

Options for Protecting Medical Data by IP Rights

Iryna Lishchuk, Marc Stauch
Institut für Rechtsinformatik
Leibniz Universität Hannover
Hannover, Germany
e-mail: lishchuk@iri.uni-hannover.de, stauch@iri.uni-hannover.de

Abstract—**This paper investigates different approaches to recognising Intellectual Property (IP) in medical data so as to protect efforts invested in enhancing the usability of such data through data curation. Sui generis database rights, copyrights and related rights, the legal regime of know-how all offer plausible options for protection here. In this paper, we analyse these options including by reference to a specific EC FP7 e-health research project (CHIC) and assess the prospects and potential benefits of applying them in order to protect investments made in data curation for medical research.**

Keywords-IP rights; data rights; medical data; data curation.

I. INTRODUCTION

In recent years attention is increasingly focusing on the potential use of clinical health data for medical research. In principle, such data, recorded in patient or research databases can be of tremendous value when analyzed, in revealing linkages, e.g., between environmental and/or genetic factors and diseases. A major advantage too is that such connections can often be identified straight from the records, without the need for further invasive and potentially risky research.

As the potential value of health data becomes better understood, efforts to monopolize clinical data by exclusive IP or proprietary rights are also expanding. For instance, there are cases when the commercial use of health related data has been asserted under the coverage of database rights [1]. Patentable inventions have been derived out of the data of patients and research subjects and successfully commercialized [2]. The property right in data, generated in medical research, may also be claimed under contractual schemes [3]. At some point copyrights may also come to consideration for monopolizing data in medical domain [4].

However, as a precondition for allowing a significant amount of clinical data to be usefully exploited, there is an important initial step required in the form of data curation. In this regard, as we analyze below, most types of IP protection are tailored to protect specific objects that have already passed a certain threshold of maturity (data repositories, confidential information with assignable commercial value, etc.); but, as we discuss, none as such guarantees adequate protection to protect the prior investment made in curating the data.

In what follows, we begin by describing the data curation process in medical research in Section II. Then, in Section III, we consider the key relevant regimes of IP protection that may apply to protect such activity, namely: copyright and related rights, sui generis data base rights, and know-how protection, as well as reliance on contractual mechanisms. By way of illustration, in Section IV, we consider how those regimes may apply to data curation in the context of a specific medical research project. In Section V the paper then concludes with some suggestions as to how curation activity may be better protected in the future.

II. DATA CURATION

The clinical data provided for e-health research usually comprises a large mass of data of multiple data types, formats, words, figures, numerical parameters, abbreviations, etc. From a technical standpoint, data integration is still a significant challenge for such research. In this regard, a starting point in the context of curation might be to see raw data in terms of the 'given', which as yet lacks semantic meaning, with the latter only emerging through the addition of an interpretive context (which also marks the change in state from data into information). It is arguably the technological development and transformation of raw or incompletely processed data into information (or the uncovering of additional semantic meaning), brought about by the curative process, which presents the suitable object of IP protection.

Data integration is key here, but the format, scope, parameter, structure, context, terminology, completeness, etc., of the individual and heterogeneous data are not standardized, which may affect their quality, and ultimately their interoperability and integration [5]. This could also potentially affect collaboration of the different researchers in this field if they use different semantics and techniques to describe, format, submit, and exchange data.

The curation required here to ensure the data relates to and measures the same phenomena with sufficient accuracy to be usable is a large and painstaking task. It includes the problem of dealing with incomplete data fields and cross-checking that various indices were measured and recorded in a similar way (e.g., images were taken using similar equipment, co-morbidities were classified using the same terminology, etc). It is evident too that considerable expertise and skill is required for it to be performed well: the curator needs to have a real feel and understanding for the subject matter in order to make sensible judgments in resolving various gaps and uncertainties.

III. Potential IP Protection

A. *Copyright and Related Rights*

Clinical data comes for the most part from clinical trials, laboratory results, medical examinations, etc. An example of the clinical data from the research project is shown in Figure 1[6]. Such data is usually expressed in some numeric parameters, figures, words, combinations of such items. The representation of clinical data in this format is suitable and useful for digital data processing. However, the isolated items, be they words, keywords, syntax, figures or mathematical concepts as such, will not attract copyright. According to the Court of Justice of European Union (CJEU), items, "*considered in isolation, are not as such an intellectual creation of the author who employs them.*" [7]. In order to be protected by copyright, the data must constitute the expression of the original author´s creativity, which is only present when "*through the choice, sequence and combination of those words that the author may express his creativity in an original manner and achieve a result which is an intellectual creation*" [7]. The protection of clinical data by copyright may be acceptable for the medical reports, written by the physician or the patient and only when the expression of original creativity is achieved.

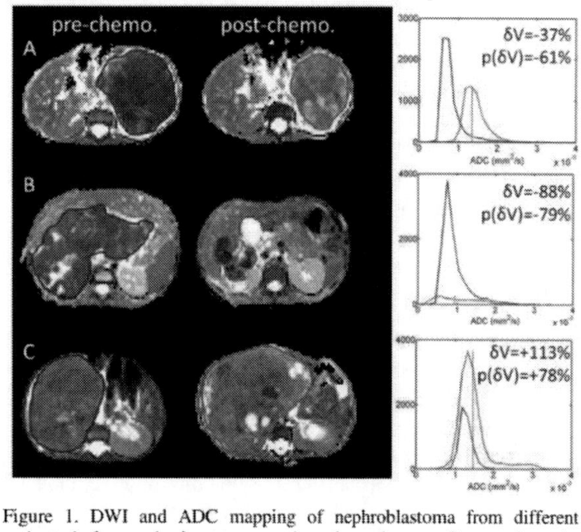

Figure 1. DWI and ADC mapping of nephroblastoma from different patients before and after pre-operative chemotherapy. Presented at the annual meeting of the British Chapter of the ISMRM, September 2012, provided by Prof. Kathy Pritchard-Jones from UCL. Copied from CHIC Deliverable D2-2 "Scenario based user needs and requirements" [6].

As may be seen from the image, some data is presented in visual form and is represented by images. However, medical images are normally produced by technical means (such as X-Ray, Ultrasound, etc.) and lack the creativity – an indispensable pre-requisite for copyright. A similar standard of copyright and requirement of original creativity applies to photographic works as well. According to Recital 16 Directive 2006/116/EC [8], a photographic work is protected by copyright, if it is original. A work "*is to be considered original if it is the author's own intellectual creation reflecting his personality*". Other criteria such as merit or purpose are not relevant for copyright. According to the CJEU decision in the case C 145/10 REC of Eva-Maria Painer [9], copyright protects pictures taken by an individual, exercising free and creative choices, thus stamping a picture with his personal touch. It means, only pictures, which are taken by an individual expressing some level of creativity may be protected by copyright. On the other hand, images, generated automatically, will lack the creative input and may not be copyrighted. Since the images, produced in medical domain, are normally taken automatically and the process of recording is mostly completely managed by technical means, such images normally do not express creativity and do not attract the protection by copyright, respectively.

Apart from the rights considered so far, in the field of copyright there are a number of other emerging rights granted as a response to relevant investment. These rights are normally provided to the person, who invests in producing the protectable information. Such rights are referred to as related rights. Protection by related rights does not necessarily link to the intellectual creation (as the case is with traditional copyright), but rather to the economic investment.

The major rationale for protection by related rights tends to shift between intellectual creation and investment [10]. A mixture of artistic creation and investment attracts exclusive rights to performers in fixations of their performances. The economic investment constitutes a major factor, which renders exclusive rights to phonogram producers in their phonograms, to the film producers in respect of first fixations of their films, to broadcasting organizations in fixations of their broadcasts [11].

However, the number of related rights as of now is rather limited (mostly to those, indicated above). Therefore, attaching added value to the data enriching, post-procession, modification, etc., does not constitute the kind of investment protectable by related rights.

Against these considerations, the protection of clinical data, which is normally collected in the course of medical examinations and is represented in some numerical or technical visual format, by copyrights or related rights, may not be considered as a practicable solution, because the requirements for copyright protection in this data would not be met.

B. *Sui Generis Database Right*

As a rule, clinical institutions, participating in medical research, manage and maintain the clinical data in the clinical data repositories. Some clinical institutions manage their clinical information and store the results of clinical trials using Ontology-based Clinical Trial Management Application (ObTiMA) [12]. Others prefer data management systems specific to their medical activities.

Against this practice, an option of protecting the clinical data under the umbrella of sui generis database rights comes into consideration first.

The legal protection of databases is provided by the Directive 96/9/EC of 11 March 1996 on the legal protection of databases (the Database Directive) [13]. Such protection is granted in recognition of the fact that constructing a database requires "*investment of considerable human, technical and financial resources*" [13]. The directive aims to reward and protect such investment by providing the maker of a database with a sui generis data base right that places him in a position to prevent unauthorized access and copying of the database contents, which he compiled. In this regard, Article 7 Database Directive states:

"*Member States shall provide for a right for the maker of a database which shows that there has been qualitatively and/or quantitatively a substantial investment in either the obtaining, verification or presentation of the contents to prevent extraction and/or re-utilization of the whole or of a substantial part, evaluated qualitatively and/or quantitatively, of the contents of that database.*" The object of protection in terms of the Database Directive is a 'database' meaning "*a collection of independent works, data or other materials arranged in a systematic or methodical way and individually accessible by electronic or other means*" [13].

Protection of databases by the sui generis right can be considered as a plausible option for protecting the clinical data repositories, provided such repositories satisfy the criteria for protection. For this, the repository must show significant investment in "*the obtaining, verification or presentation*" of its contents.

As regards the scope of the database right, it would protect the collected data from being copied as a whole or in substantial part, evaluated "*qualitatively and/or quantitatively*" and either copied in one action or step by step [13].

Provided the clinical data repository qualifies as a database in the meaning of Database Directive and the clinical institution holds the sui generis database rights, the institution may stipulate the terms of using the repository contents as a whole, grant the rights of use under contractual license, prevent and enforce the unauthorized extraction/reutilization of the repository contents as a whole or in substantial part. The holder of sui generis database rights may leverage how the contents of its repository may be used, whether the data items may be extracted (downloaded) and in what scope, whether the data may be transferred to external parties or whether the data procession may only be done on its premises.

However, the sui generis protection applies to the contents of the repository as a whole or in substantial part and may apply separately and irrespective of protectability of data items by other rights, such as copyrights. Article 7 (4) makes this explicit, saying that the database right: "*shall apply irrespective of eligibility of the contents of that database for protection by copyright or by other rights. Protection of databases [....] shall be without prejudice to rights existing in respect of their contents*".

Thus, the holder of the repository may manage the use of the repository contents as a whole. However, the use of separate data items in the repository may remain governed by the terms, stipulated by the data providers and/or holders of rights in such items. For instance, the access rights to the datasets, handled as confidential, may require signing of non-disclosure agreement (NDA) and the use of such data may be limited and be subject to technical protection measures, etc.

The options of protection, which potentially may apply to separate datasets we consider next.

C. *Know-how*

Because of the high sensitivity of health related data (and the potential harm from disclosure to the patient's interests in privacy, dignity and autonomy), clinical data in the medical treatment domain is managed under the rules of professional medical secrecy and subject to the fiduciary duties. For preserving the secrecy of clinical data, after such data leaves the medical domain (where it was handled under the rules of professional medical secrecy) and enters the domain of clinical research (where not necessarily all parties are bound by the rules of professional secrecy), protecting such data under the legal regime of know-how (or as undisclosed information) may be advised as a good option.

Protection of undisclosed information is provided by Section 7, Article 39 et seq. TRIPS Agreement [14]. The legal regime of know-how enables natural and legal persons, who are in legitimate possession of such information, to prevent such information "*from being disclosed to, acquired by, or used by others without their consent in a manner contrary to honest commercial practices.*" Unfair practices for these purposes would include the acquisition of information via violation of contractual duties, breach of confidentiality obligations, inducement to breach, etc. [14].

In order to be protectable, the relevant information should have the quality of protectable information within the meaning of Article 39 TRIPS Agreement. Article 39 TRIPS Agreement protects information, which:

"*(a) is secret in the sense that it is not, as a body or in the precise configuration and assembly of its components, generally known among or readily accessible to persons within the circles that normally deal with the kind of information in question;*

(b) has commercial value because it is secret; and

(c) has been subject to reasonable steps under the circumstances, by the person lawfully in control of the information, to keep it secret." [14].

The first weak point of protecting clinical data as know-how is that as of now the legal framework on know-how protection in the EU is not harmonized [15]. Although, there is a proposal for a draft directive on the protection of undisclosed information in the EU (the Draft Directive)

[16], before it is adopted and implemented, protection of know how remains dispersed through the national states of the EU Member States, and subject to varying requirements for and scope of protection

The Draft Directive, which is intended to harmonize the national laws in relation to know-how protection, in many aspects repeats the provisions of the TRIPS Agreement (in particular, it relates to the protectable subject matter and requirements for protection (Article 2), acts of unfair acquisition of information (Article 3), rights and remedies conferred (Article 5 et seq), etc.). In this regard it may also be queried how far the Draft Directive, if adopted, would improve the protection for data, the preparation of which consumed much effort, but which for one or another reason may not reach the level of protectable know-how. Here the key obstacles in applying know-how protection to the clinical data, processed for research, relate to the need (in order to be protected) for such data to be secret, subject to the confidentiality measures and have economic value.

First, to satisfy the criterion of secrecy, the information, sought to be protected, must be accessible to a limited number of persons only. The use of such information must be subject to confidentiality measures. The application of confidentiality measures means that the data must be stamped as "Confidential" and the sharing of such data must be made upon non-disclosure obligation and observation of the confidentiality measures. Disclosure of such datasets without due confidentiality measures might compromise the regime of secrecy so that protection would be forfeited. As regards the requirement of economic value of know-how, this will be considered to be present if through publication, the research investment and competitive standing of the entity doing the work would be undermined [17].

In relation to the volumes of clinical data, made available for research, this requirement, besides being at odds with the underlying culture of academic research, would create further workload. The data, subject to the regime of confidentiality, must first be strictly identified. The confidentiality mark would need to be attached to individual data items and any use and disclosure of such data to any third party must be made upon signing the non-disclosure agreement. This preservation of the confidentiality mark, conclusion of NDA and control over handling such data as confidential would present another challenge.

Against these considerations, the protection of clinical data under the legal regime of know-how might, in principle, be possible in relation to some defined amount of data, but hardly offers a feasible solution, when protection of large amounts of data, processed in medical research is sought. It also may operate against the principle of openness, if optimal use is to be made of the data by the research community, exploiting the full potential of available datasets.

D. *Contractual Approaches*

Insofar as the IP regimes for protecting the data, produced in medical research projects fail, one further method for regulating rights in data may be by contractual relations. Thus, in third party funded projects, the relations of ownership over the research results are typically governed by contract. The sponsor is typically interested to exploit the project results and funding is typically granted upon condition that the sponsor acquires the ownership and exploitation rights over the research results [3]. This model does not cause problems in practice, because the acquisition of ownership and exploitation rights is typically foreseen by the contract. The participating institutions are bound by these contractual relations and required to procure the ownership over the research results from the persons, whom they engage into the project.

IV. Application of IP Regimes to Data Curation in CHIC

A. *Background*

The research project "Computational Horizons In Cancer (CHIC): Developing Meta- and Hyper-Multiscale Models and Repositories for In Silico Oncology", is an ICT research project in the clinical domain [18]. CHIC develops clinical trial driven tools and services within a secure infrastructure, which facilitate the creation of multiscale cancer hyper-models (integrative models) by technical means. These composite multiscale constructs of models (hyper-models or integrative models) are intended to synthesize and imitate the biological processes, which occur in course of tumor progression, at several temporal and spatial levels (molecular, cellular, etc.) at once.

In this context too, the study of how individual cancer components interact with each other has led to an explosion in the number of different types of data generated from the patients such as: molecular data, epigenetic data, clinical data, imaging data, pathology data and other laboratory data [19]. These different data types are assembled in order to systematically explore and formalize them in mathematical models.

Subsequently, the models are developed and validated against clinical data either taken from the literature or provided by the clinical partners [20]. The data management systems, used by the clinical partners, differ. Whereas the integration of data from data management system ObTiMA [12] is harmonized, the data from individual clinical data repositories need to be adapted to the requirements of the project. The use of diverging data management systems by the clinical institutions leads to the situation that the data, collected from different sources, is not inter-operable with each other and mostly cannot be used for research as such. The clinical data also needs to be post-processed by the modelers so that it fits into the set of parameters, which the models recognize and can utilize as an input for running the simulations. Data curation is a very important step because the inputs, outputs and descriptions of processes, simulated by the models, need to be standardized into the set of parameters, acceptable and usable by all cancer models.

B. Applicability of IP Regimes to Project Data Curation

The clinical data, which after the necessary de-identification enters the domain of CHIC, is placed and stored in the CHIC clinical data repository. The CHIC data repository hosts data categorized per data type: imaging data (DICOM etc), descriptive/structural data (age, sex etc), other files (histological reports), links (to other data repositories) etc. The datasets for each type are accessible individually so that the data corresponding to the model parameters may be chosen. The fact that the repository is built "*based on the experience already accumulated during the implementation of other data repositories*" should be sufficient to prove the requisite investment in "*either the obtaining, verification or presentation*" of its contents [13]. Against this background, the database right in the CHIC clinical data repository is likely to be granted.

Protection of the CHIC data repository by the sui generis database rights would go to the maker of the database. In the meaning of the Database Directive, the maker of a database is seen as "*the person who takes the initiative and the risk of investing*", but excluding subcontractors [13]. Thus, the party, who constructed the CHIC repository, would be in a position to manage the use of the repository, such as by allocating the access rights to the project parties or external parties, to define the rights of use (access only, modification, download, etc.), to divide the repository into sections and define different regimes of uses depending on the data stored therein, etc. Grant of the sui generis protection would also entitle the right holder to enforce his rights, once unauthorized copying of the repository contents on the large scale has occurred.

Apart from the protection of the repository contents as a whole by sui generis database rights, the items in the repository may also enjoy protection in their own right. Since the clinical data repository deals with highly sensitive information (meaning that already for that reason, access to the data is strictly limited), application of the legal regime of know-how to some data items at least may be an option. As we saw above, for this, the data items, selected for know-how protection, must be identified, the access and use of such data shall be limited to a defined number of people only, the management of such data shall be subject to confidentiality measures. In the case of CHIC, the regime of secrecy may be provided to the data via marking it as "Confidential" and making the disclosure of such data subject to the non-disclosure obligation. Considering from the technical side, the confidentiality mark would then need to be placed and borne by the data throughout the whole research process so that the data marked as "confidential" by the input comes out marked "confidential" by the output. This would present an additional workload, but is implementable. Also, disclosure of such data items to the CHIC parties subject to the non-disclosure obligation would not present a significant obstacle, because the project parties are bound by the contractual relations within the project. The factual use of data within the project may also be managed by technical measures, such as granting or denying the access rights, rights of use and extraction, limiting the data procession to the framework of technical infrastructure of CHIC only. Whereas the application of such contractual and technical confidentiality measures to the clinical data in CHIC may be feasible, in how far such technical and confidentiality measures may be implemented in other medical research projects may be questionable.

By contrast, copyrights and related rights offer less plausible options for protecting the clinical data in CHIC. As noted above, the clinical data in CHIC is represented by technical data from clinical trials, which is composed from different parameters. As observed in Section III, isolated items are not protectable by copyright. Copyright will fail against the lack of creativity expressed in such data. The investment, deployed in curating the data for CHIC, does not qualify as investment, protectable by related rights.

V. CONCLUSIONS

As we have seen, there are various ways in which the activity of curating clinical datasets could benefit from IP protection. Thus, collecting, arranging the data into a repository and making it suitable for use may render the investment, deployed in collecting and presenting the data, protectable by sui generis database rights. Similarly, the generation of research data and adoption of additional confidentiality and security measures to keep this data secret to the broader community may render such data protectable as know-how.

However, the present approach that seeks to maintain (commercial) data confidentiality by keeping data secret leads to a fragmented research environment, and reduces the chances for greater data interoperability to be achieved. Here the law - aided by technology should aim to encourage greater openness, while assuring appropriate curation rewards. This could, e.g., take the form of an officially endorsed mechanism or system for measuring and tagging changes produced in a given data set (or the merging of several data sets) resulting from curation efforts, as the reward-trigger. At the same time, as another crucial policy element, the law needs – especially in the case of the curation of sensitive health data – to ensure that privacy and other interests of patients and research subjects are and remain adequately protected.

In particular, it will here be necessary to take account of (and compensate for) the knock-on effects of IP changes, where data-holders are no longer (also) motivated by commercial considerations to keep their data secure and confidential. This concern is all the greater here since the activities of data sharing and curation being encouraged, also by their nature present enhanced risks to personal privacy. The point of curation is precisely to uncover new connections and patterns in data that help generate robust inferences (usable – for good or ill) about the relevant data subjects. Accordingly, it is submitted that any system for rewarding investment in data curation should also require (as a condition for such rewards) that the data curator takes every appropriate measure to counterbalance the associated enhanced risks to privacy.

ACKNOWLEDGMENT

The research leading to these results has received funding from the European Union Seventh Framework Programme FP7/2007-2013 under grant agreement No 600841.

REFERENCES

[1] D. M. Gitter, "Informed Consent and Privacy of De-Identified Information and Estimated Data, Lessons from Iceland and the United States in an Era of Computational Genomics," (Published Conference Proceedings style), in Proc. ALLDATA 2016, The Second International Conference on Big Data, Small Data, Linked Data and Open Data (includes KESA 2016), Lissabon, 2016, pp. 7-12.

[2] Moore v. Regents of University of California, Supreme Court of California, July 9, 1990, 51 Cal. 3d 120.

[3] H.-D. Lippert, „Wem gehören Daten, die im Rahmen von Forschungsprojekten gewonnen werden?"/"To whom belongs the data generated in research projects?" in: Geistiges Eigentum: Schutzrecht oder Ausbeutungstitel?/in: Intellectual Property: Protection right or title to exploit, Springer, Volume 5, 2008, pp. 359-369.

[4] Haimo Schack, "Zur Rechtfertigung des Urheberrechts als Ausschliesslichkeitsrecht"/"On justification of copyright as exclusive right," in: Geistiges Eigentum: Schutzrecht oder Ausbeutungstitel?/in: Intellectual Property: Protection right or title to exploit, Springer, Volume 5, 2008, pp. 124-140.

[5] European Commission, "E-Health Task Force Report – Redesigning health in Europe for 2020," Luxembourg 2012, ISBN 978-92-79-23542-9

[6] CHIC, Deliverable D2-2 "Scenario based user needs and requirements", <http://chic-vph.eu/uploads/media/D2-2_Scenario-based_user_needs_and_requirements.pdf> 16.08.2016.

[7] CJEU, Judgment of 16 July 2009, Case C 5/08, Infopaq International A/S v Danske Dagblades Forening, Ref. 45.

[8] Directive 2006/116/EC on the term of protection of copyright and certain related rights (codified version), OJEU, L 372/12, 27 December 2006.

[9] CJEU, Judgment of 7 March 2013, Case C 145/10 REC, Eva-Maria Painer v. Standard VerlagsGmbH, Axel Springer AG, Süddeutsche Zeitung GmbH, Spiegel-Verlag Rudolf Augstein GmbH & Co. KG, Verlag M. DuMont Schauberg Expedition der Kölnischen Zeitung GmbH & Co. KG.

[10] H. Zech, "Information als Schutzgegenstand," Tübingen, 2012, ISSN: 0940-9610 (Jus Privatum).

[11] Directive 2001/29/EC of 22 May 2001 on the harmonisation of certain aspects of copyright and related rights in the information society, OJEU L 167/10 - L 167/19, 22.6.2001.

[12] H. Stenzhorn, et al, "The ObTiMA system - ontology-based managing of clinical trials," Stud Health Technol Inform. 2010;160(Pt 2):1090-4.

[13] Directive 96/9/EC of the European Parliament and of the Council of 11 March 1996 on the legal protection of databases, OJEU, L 77/20 - L 77/28, 27.3.96.

[14] Agreement on Trade-Related Aspects of Intellectual Property Rights, the TRIPS Agreement, Annex 1C of the Marrakesh Agreement Establishing the World Trade Organization, Marrakesh, Morocco, 15 April 1994.

[15] Hogan Lovells International LLP, "Report on Trade Secrets for the European Commission – Study on Trade Secrets and Parasitic Copying (Look-alikes), MARKT/2010/20/D", 2011.

[16] Proposal for a Directive on the protection of undisclosed know-how and business information (trade secrets) against their unlawful acquisition, use and disclosure, Brussels, 28.11.2013, COM(2013) 813 final, 2013/0402 (COD).

[17] K. Lodigkeit, Intellectual Property Rights in Computer Programs in the USA and Germany, Peter Lang GmbH, 2006, pp. 98-101.

[18] Computational Horizons In Cancer (CHIC): Developing Meta- and Hyper-Multiscale Models and Repositories for In Silico Oncology <http://chic-vph.eu/project/> 16.08.2016.

[19] C. Coveney, J. Gabe, and S. Williams, "The sociology of cognitive enhancement: medicalisation and beyond," Health Sociol. Rev., 20 (2011), pp. 381–393.

[20] J. Dejaegher, L. Solie, S. De Vleeschouwer, and S. W. Van Gool, "Dendritic Cell Vaccination for Glioblastoma Multiforme: Clinical Experience and Future Directions," In G. Stamatakos and D. Dionysiou (Eds): Proc. 2014 6th Int. Adv. Res. Workshop on In Silico Oncology and Cancer Investigation – The CHIC Project Workshop (IARWISOCI), Athens, Greece, Nov.3-4, 2014 (www.6thiarwisoci.iccs.ntua.gr), pp.14-18. (open-access version), ISBN: 978-618-80348-1-5.

Effect of SinR Transcriptional Factor on the Expression of Bacilysin Biosynthetic Operon in *Bacillus subtilis*

Ayten Yazgan-Karataş; Öykü İrigül-Sönmez
Molecular Biology-Biotechnology and Genetics Research Center (MOBGAM) and Molecular Biology and Genetics Department, Istanbul Technical University
Istanbul,Turkey
e-mails: karatasay@itu.edu.tr; oirigul@iontek.com.tr

Hasan Demirci
Molecular Biology and Genetics Department, Yıldız Technical University
Istanbul, Turkey
e-mail: hasan@yildiz.edu.tr

Abstract— **In *Bacillus subtilis*, bacilysin is a nonribosomally synthesized dipeptide antibiotic composed of L-alanine and L-anticapsin. It is active against a wide range of bacteria and even *Candida albicans*. The biosynthesis of bacilysin depends on the *bacABCDEywfG* operon (*bac* operon) and the adjacent monocistronic gene *ywfH*. In our previous study, *lutR* mutation significantly decreased the maximum transcription level of the bacABCDEywfG operon at the onset of stationary phase to about 57% of wild-type level. In this study, we aimed to test the possible effect of global regulator SinR on the expression of the *bac* operon. For this *sinR* gene was disrupted in the transcriptional *bacA-lacZ* fusion bearing strain (OGU1) to generate "*sinR::cm bacA::lacZ::erm*" bearing strain (OGU1SR). Additionally, we also test whether *sinR* and *lutR* gene products affect the *bac* operon expression mutually due to the close regulatory interactions between LutR and SinR transcriptional factors . For this, *ΔsinR-ΔlutR* double mutant (*sinR*::cm *lutR::Tn10::spc bacA::lacZ::erm*) strain (OGU1SRLR) was constructed. Finally, all of that resulting strains and OGU1 were grown in Perry and Abraham (PA) medium at 37°C and β–galactosidase activities were measured throughout the different stages of growth. β–galactosidase assay results indicated that in the absence of *sinR* gene product *bac* operon expression was severely effected. Since, disruption of *sinR* gene resulted with complete loss of transition state dependent induction of *bac* operon expression while almost the same *bacA*-expression profile as the single *sinR* mutant was detected in the *sinR-lutR* double mutant.**

***Keywords- Bacillus subtilis;* bacilysin*; bacABCDEywfG; bac* Operon**

I. INTRODUCTION

Antibiotics have a crucial role in keeping the public healthy. Understanding the genetic and molecular basis of the regulation of antibiotic biosynthesis is of great interest for providing new strategies for targeted genetic engineering of antibiotic producing strains.

Bacilysin is a simple dipeptide antibiotic which is composed of an L-alanine residue at the N-terminus and an unusual amino acid, L-anticapsin, at the C-terminus [1] Antimicrobial activity of bacilysin depends on the anticapsin moiety [2]. It is active against wide range of bacteria and even *Candida albicans*.

Bacilysin biosynthesis is regulated either positively or negatively. Global regulatory proteins CodY and AbrB negatively regulate the *bac* operon, on the other hand global regulatory proteins Spo0A and ComA positively regulate the *bac* operon expression [3]. In addition to master regulatory genes, bacilysin biosynthesis in *Bacillus subtilis* was also demonstrated to be under the control of GntR type transcriptional regulator LutR which was identified by transposon mutagenesis technique [4].

SinR is a global transcriptional regulator that is required for sporulation, competence, motility, exoprotease production and biofilm formation in *B. subtilis*. In our previous study, it was shown that LutR is a transcriptional factor that has a function in the regulation of many pathways and there is a close overlap among the targets of LutR and SinR regulatory proteins [5]. In this study, we aimed to test the possible effect of global regulator SinR on the expression of the *bac* operon as well as to test whether *sinR* and *lutR* gene products affect the *bac* operon expression mutually by using *lacZ* fusion analysis.

The rest of this paper is organized as follows. In Section II we described the materials and methods used in this study. In Section III we described and evaluated the main results obtained in this study.

II. MATERIALS AND METHODS

A. *Bacterial strains, media and culture conditions*

The strains of *B. subtilis* used in this study are listed in Table 1. All these strains were the derivatives of *B. subtilis* 168.

TABLE 1. BACTERIAL STRAINS AND THEIR GENOTYPE

Strain	Genotype	Source
B.subtilis PY79	Wild type, BSP cured prototrophic derivative of *B.subtilis* 168	P.Youngman
OGU1	*bacA::LacZ::erm*	İ.Öğülür
SinR mutant *B. subtilis*	*ΔsinR::cm*	AkosT. Kovacs
OGU1LR	*lutR::Tn10::spc bacA::lacZ::erm*	T. E. Köroğlu
OGU1SR	*ΔsinR::cm bacA::LacZ::erm*	In this study
OGU1SRLR	*lutR::Tn10::spcΔsinR::cm bacA::lacZ::erm*	In this study

B. subtilis strains were normally grown in Luria Bertani (LB) medium at 37°C. The Perry and Abraham (PA) medium was used for bacilysin production. Antibiotic concentrations employed in this study for direct selection were as follows: erythromycin (1 µg/ml), lincomycin (25µg/ml), spectinomycin (100 µg/ml), chloramphenicol (5 µg/ml) and ampicillin (100 µg/ml).

B. *Insertional inactivation of regulatory genes*

For construction of OGU1SR (*sinR::cm bacA::lacZ::erm*) strain , genomic DNA of *sinR* disrupted *B. subtilis* strain that harbors *ΔsinR::cm* mutation was isolated and transformed into the competent cells of *bacA-lacZ* fusion bearing strain OGU1. *ΔsinR-ΔlutR* double mutant strain OGU1SRLR was constructed by transforming competent cells of OGU1LR (*lutR::Tn10::spc bacA::lacZ::erm*) strain with chromosomal DNA from OGU1SR (*sinR::cm bacA::lacZ::erm*) strain.

C. *β–Galactosidase Assay*

β-Galactosidase assay was carried out as described by Miller (1972). The specific activitiy is expressed as [A_{420}-1.75 X A_{550} per Minute per OD_{595}]X1000.

III. Results and Discussion

To analyse the effect of *sinR* and *lutR* insertional inactivations on the *bac* operon expression, transcriptional *lacZ* fusion analysis is performed. For that purpose, we first constructed the congenic derivatives of the transcriptional *bacA-lacZ* fusion bearing OGU1 with the mutation of the interested genes *sinR* and *lutR*. The resulting strains OGU1SR, OGU1SRLR, OGU1LR and OGU1 as a control, were cultured in PA medium at 37°C and sampled in every 1 h for the β-galactosidase assay. As shown in Figure 1, expression of the *bac* operon was persistent during the exponential growth phase. However, it induced in the course of transition between exponential and stationary phases and reached to its top level upon entry into stationary phase. As shown previously, *lutR* mutation significantly reduced the maximum transcription level of the *bac* operon (Figure 1). On the other hand, the deletion of the *sinR* gene severely affected the *bacA-lacZ* expression and resulted with the complete loss of transition state dependent induction of the *bac* operon expression. However, almost the same *bacA*-expression profile as the single *sinR* mutant was detected in the *sinR-lutR* double mutant.

IV. Conclusion

Results of this study indicate that both SinR and LutR regulatory proteins are required for the maximum *bac* operon expression but SinR activity is essential for the transition state induction of the *bac* operon expression.

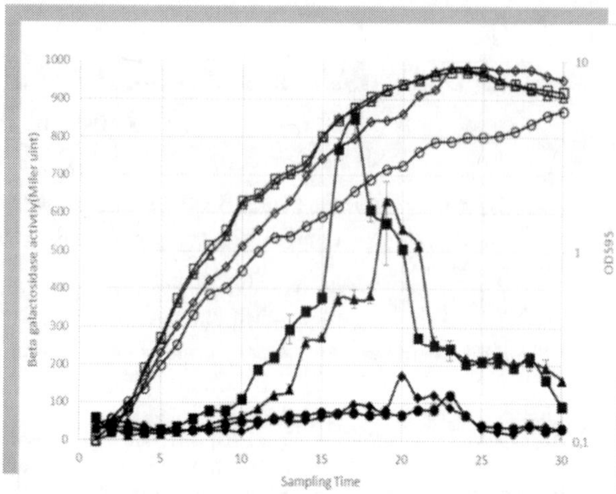

Figure 1. Growth curves and β-galactosidase activity of *bacA::lacZ* fusion OGU1 (white squares and black squares) and its derivatives *lutR* mutant (white triangles and black triangles), *sinR* mutant (white circles and black circles) and *lutR-sinR* mutant (white diomands and black diomands) in PA medium. Error bars represent the standard deviation of three independent experiments .

Acknowledgment

This work was supported by Istanbul Technical University Scientific Research Foundation.

References

[1] J. E Walker and E. P Abraham., "the Structure of Bacilysin and Other Products of *Bacillus subtillis*". Biochem. J. vol.118, pp.563-570,1970.

[2] M. Kenig, E. T. Vandamme and E. P.Abraham, "the Mode of Action of Bacilysin and Anticapsin and Biochemical Properties of Bacilysin-Resistant Mutants." Microbiol. vol. 94, pp. 46-54, 1976.

[3] T. E. Köroğlu, Ö. İsmail, S. Mutlu, A. Yazgan-Karataş and G. Özcengiz, "Global regulatory systems operating in bacilysin biosynthesis in *Bacillus subtilis*." J. Mol. Microbiol. vol. 20, pp. 144-155,2011.

[4] T.E. Köroğlu, G. Kurt-Gür, E. C. Ünlü and A. Yazgan-Karataş, "the Novel Gene *yvfI* in *Bacillus subtilis* is Essential for Bacilysin Biosynthesis." Antonie van Leeuwenhoek, vol. 94, pp. 471-479, 2008.

[5] Ö. İrigül-Sönmez, T.E. Köroğlu. B. Öztürk, A. T. Kovacs, O. P. Kuipers and A. Yazgan-Karataş, "in *Bacillus subtilis LutR* is Part of the Global Complex Regulatory Network Governing the Adaptation to the Transition from Exponential Growth to Stationary phase." Microbiol. vol.160, pp. 243-260, 2014.

[6] j. H. Miller J.H. "Experiments in Molecular Genetics. Cold Spring Harbor, Cold Spring Harbor Laboratory,1972.

Using a Dual Vibration Absorber to Suppress Rest Hand Tremor of Elderly

Sarah Gebai, Mohamad Hammoud, and Hassan Khachfe
Biomedical and Mechanical Engineering Departments
Lebanese International University,
Beirut, Lebanon
{mohamad.hammoud, hassan.khachfe}@liu.edu.lb
sarah.gebai@gmail.com

Abstract—**Human tremor is a public health problem that can lead to social and physical deterioration. Parkinson disease (PD) is a slowly progressive degenerative disorder of the central nervous system. It is more significant in elderly patients. Its signs start to appear at the age of 59 years. Levodopa is the most commonly used medication to reduce tremor in PD patients, but has serious side effects. Vibration absorbers can be used as a mechanical treatment that can counter-act the involuntary tremor caused by difficulties to give certain commands to the muscles. The hand is modeled at the musculoskeletal level as a three degree-of-freedom (DOF) system in the horizontal plane. The vibration absorbers are designed to reduce the flexion angle at the shoulder, elbow and wrist joints when the muscles are operating at the resonance frequencies that belong to the range of resting tremor. Single and dual series absorbers are designed to satisfy the tuning conditions. The system's behavior is analyzed in the frequency and time domains. The suggested dual absorber is composed of a series combination between the elastic absorber and the viscous damper absorber. It is more efficient than a single absorber of the same total mass and causes 85.7–86.6%, 88.4–91.3% and 58.29–59.9% amplitude reduction at the shoulder, elbow and wrist joints, respectively.**

Keywords-Parkinson disease; elderly; tremor reduction; mechanical treatment; dual vibration absorber.

I. Introduction

Parkinson disease (PD) is a multi-system neurodegenerative disorder caused by a deficiency of neurotransmitter dopamine within the brain and affects the brain's control of the muscles. It leads to shaking (tremor), increased muscle tone (stiffness), slowed movements and balance problems. Parkinson tremor is usually a resting tremor characterized by adduction-abduction or flexion-extension motion with frequency 3–7 Hz [1]. PD incidence increases with aging. It is difficult to specify the exact number of young patients affected by PD, but we know that they are rarely affected by this illness. It is more significant in elderly patients and its average onset age is 59 years [2]. Initial symptoms may start early in life, but tremor progresses over time and becomes significant to a physician when the patient is elderly.

Dopaminergic drugs are medications which aim to temporarily restore correct dopamine level in the substantia nigra and striatum. This treatment reduces the motor symptoms and signs of PD, without, however, curing the disease or stopping its progression. Levodopa is the most effective dopaminergic medication for elderly patients which is converted in the brain into dopamine. The recommended drug dose increases with the increased symptom severity, but high doses can produce involuntary tremor and can lead to serious side effects. In addition, about 25% of PD patients can have a low quality of life since they do not respond to drugs or neurosurgery treatments [3][4].

Vibration absorbers can be used as a mechanical treatment to reduce the tremor of elderly Parkinson patients by supplying certain commands to muscles to counter-act the vibration. Hand's involuntary tremor is transmitted to the absorber, which causes the vibration of its proof mass. Mechanical oscillations are considered as the main source for the tremor, in which joints and muscle movements satisfy the laws of physics. Therefore, a dynamic model of the human hand can describe its response and can be used for the numerical testing of absorber's performance.

A method of biodynamic response has recently been used to describe the motion of human hand and solve problems related to tremor. Jackson et al. [5] modeled human upper limbs as two pivoted straight rods (the upper arm, and the forearm together with the hand) with concentrated mass at centroid of each segment. The hand was modeled in vertical plane during locomotion where flexion-extension planar motion at the joints was considered. The described model shows similar results to the available data on the movement of the upper limb during locomotion. Raikova [6] described the real anatomy of the muscle functions by modeling the human upper limb as a seven degree-of-freedom (DOF) system. The biomechanical model consists of the upper arm, forearm and palm modeled as rigid bodies connected by the shoulder joint (three-DOF spherical joint-connected to immobile trunk), elbow joint (two one-DOF pin joints) and wrist joint (two-DOF Hook's joint).

Hashemi et al. [7] designed a single DOF vibration absorber that is and attached to the forearm. It was able to reduce tremor amplitude numerically for the model hand and experimentally for the fabricated model. Igusa and Xu [8] studied the multiple mass dampers tuned within a frequency range. They found that it is more robust than the tuned mass damper with same the total mass. Gebai et al. [9] have designed a single dynamic vibration absorber (DVA) capable to suppress the steady state response of the hand system excited at its fundamental frequency. Moreover, they proposed a single DVA which was able to help elderly patients suffering from neurodegenerative disorder by reducing the homogenous as well as the steady state response of the involuntary tremor at the hand joints [10]. They have also suggested a dual parallel DVA to reduce the pathological tremor in the hands of Parkinson patients when excited at the dual harmonic resonance frequencies [11][12]. The study reveals that the dual DVA was more effective than the single DVA with the same total mass.

In this study, a new three-DOF hand system is used to describe the biodynamic response of the hand of a PD Patient. The system is operating at the first two natural frequencies due to the shoulder, elbow and wrist muscles activation. The dynamic model of the hand is used to study the performance of a dual DVA in suppressing the rest tremor of a PD patient. This absorber is formed from an elastic absorber connected in series to a viscous damper absorber; we will call it the dual series elastic-viscous damper (SEVD) absorber. This absorber is connected to the forearm of the hand and compared to the single TVA of the same total mass. Both absorbers are designed to satisfy the tuning condition. The equation of motion for the system is derived using the non-Lagrangian formulation and solved using system's transfer function. The flexion angle at the hand joints are shown in the frequency and time domains.

The paper is organized as follows: Section 2 presents the dynamic structure of the model hand system, its sized segments and the derived equations of motion. Section 3 describes the designing steps for tuning the absorbers. Section 4 presents the frequency and time domain response at hand's proximal joints. Section 5 provides a comparison with the previous studies done. Section 6 includes the conclusions and recommendation for future work.

II. Biodynamic Hand Modeling

The human hand shown in Fig. 1 is modeled in the horizontal plane as a three-DOF system reflecting the biodynamic response of a Parkinson Patient. Most researchers agreed on modeling the bones and corresponding soft tissues as rigid bodies connected by frictionless joints with fixed axes or centers of rotation [13]. So, the hand segments are described as rigid bodies where the upper arm and forearm (ulna and radius) are modeled as truncated cones and the palm as rectangular plates to specify their volume (V). The three segments are connected by one-DOF frictionless kinematic pairs to permit the flexion-extension planar motion at the proximal joints. The upper arm is pinned by the shoulder joint, which is fixed to the trunk in order to reflect the resting condition of the hand. The length (l), position of centroid from proximal joints ($\bar{r}$) and density (D) used for each segment are provided experimentally by Drillis et al. [14] and shown in Table 1. Then, the mass (m) and the mass moment (I) of each sized segment are calculated using the following equation:

$$m = D.V, \quad \bar{r} = \frac{\int r^2 dm}{\int dm} \quad and \quad I = \int \bar{r}^2 dm \tag{1}$$

and the obtained results are listed in Table 2. The four muscles modeled to produce a movement are: the single joint shoulder, elbow and wrist joint muscles and the Biceps brachii muscle. Theoretically, muscles can be assumed to be regulated independently [7]. The modeled stiffness and damping coefficients of the considered muscles are provided in Table 3. Damping and stiffness coefficients of the muscles are assumed to be linearly proportional [15]. The active inputs can be described as muscular activity [16] and can be considered as sinusoidal function(s) [17].

Equation of motion for the hand system with a single DVA attached to the forearm is provided by Gebai et al. [9][12] and derived using the non-Lagrangian formulation. A similar strategy is used for the hand controlled by the dual SEVD DVA, where the general equation of motion is:

$$[M]\{\ddot{\theta}\} + [C]\{\dot{\theta}\} + [K]\{\theta\} = \{f\} \tag{2}$$

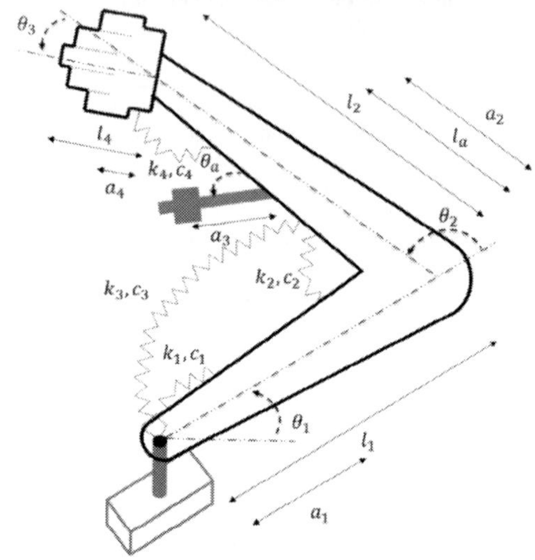

Figure 1. Dynamic model of human hand at the musculoskeletal level.

TABLE I. Collected Hand Arm Parameters [14]

Right Hand	Length (cm)		Centroid (m)		Density (kg/m^3)	
Upper arm	l_1	36.4	a_1	0.427 l_1	D_1	2.070
Forearm	l_2	29.9	a_2	0.417 l_2	D_2	1.160
Palm	l_4	20.3	a_4	0.361 l_4	D_4	0.540

TABLE II. Calculated Hand Arm Parameters

Right Hand	Upper arm		Forearm		Palm	
Mass (kg)	m_1	2.070	m_2	1.160	m_4	0.540
Inertia ($kg.m^2/rd$)	I_1	0.0228	I_2	0.0082	I_4	0.0012

TABLE III. Designed Parameters of Joint's Muscles

Muscle	Shoulder	Elbow	Biceps	Wrist
$k(N.m/rd)$	180	70	40	10
$c(N.m.s/rd)$	0.002 k_1	0.002 k_2	0.002 k_3	0.002 k_4

$$\theta = \{\theta_1 \quad \theta_2 \quad \theta_2 \quad \theta_{a_1} \quad \theta_{a_2}\}^T \tag{3}$$

$$\text{Where,} \quad M = \begin{bmatrix} M_{11} & M_{12} & M_{13} & M_{14} & M_{15} \\ M_{21} & M_{22} & M_{23} & M_{24} & M_{25} \\ M_{31} & M_{32} & M_{33} & M_{34} & M_{35} \\ M_{41} & M_{42} & M_{43} & M_{44} & M_{45} \\ M_{51} & M_{52} & M_{53} & M_{54} & M_{55} \end{bmatrix} \tag{4}$$

$$M_{11} = \left(I_1 + m_1 a_1^2\right) + \left(I_2 + m_2 a_2^2\right) + m_2 l_1^2 + m_4\left(l_1^2 + (l_2 + a_4)^2\right) + m_{a_1}\left(l_1^2 + (l_a + a_3)^2\right) + m_{a_2}\left(l_1^2 + (l_a + a_3 + a_5)^2\right)$$

$$M_{12} = \left(I_2 + m_2 a_2^2\right) + m_4 (l_2 + a_4)^2 + m_{a_1}\left(l_a^2 + a_3^2 + 2 l_a a_3\right) + m_{a_2}(l_a + a_3 + a_5)^2, \quad M_{13} = m_4\left(a_4^2 + l_2 a_4\right)$$

$$M_{14} = m_{a_1}\left(a_3^2 + 2 l_a a_3\right) + m_{a_2}(l_a + a_3 + a_5)(a_3 + a_5)$$

$$M_{15} = m_{a_2}\left(l_a a_5 + a_3 a_5 + a_5^2\right), \; M_{21} = M_{12}, \; M_{22} = M_{12}$$

$$M_{23} = M_{13}, \; M_{24} = M_{14}, \; M_{25} = M_{15}, \; M_{31} = M_{13}$$

$$M_{32} = M_{23}, \; M_{33} = I_4 + m_4 a_4^2, \; M_{34} = 0, \; M_{35} = 0$$

$$M_{41} = M_{14}, \; M_{42} = M_{24}, \; M_{43} = 0$$

$$M_{44} = m_{a_1} a_3^2 + m_{a_2} (a_3 + a_5)^2$$

$$M_{45} = m_{a_2}\left(a_3 a_5 + a_5^2\right), \; M_{51} = M_{15}, \; M_{52} = M_{25} \; .$$

$$M_{53} = 0, M_{54} = M_{45}, \; M_{55} = m_{a_2} a_5^2$$

$$K = \begin{bmatrix} k_1 + k_3 & k_3 & 0 & 0 & 0 \\ k_3 & k_2 + k_3 & 0 & 0 & 0 \\ 0 & 0 & k_4 & 0 & 0 \\ 0 & 0 & 0 & k_a & 0 \\ 0 & 0 & 0 & 0 & 0 \end{bmatrix} \tag{5}$$

$$C = \begin{bmatrix} c_1 + c_3 & c_3 & 0 & 0 & 0 \\ c_3 & c_2 + c_3 & 0 & 0 & 0 \\ 0 & 0 & c_4 & 0 & 0 \\ 0 & 0 & 0 & 0 & 0 \\ 0 & 0 & 0 & 0 & c_a \end{bmatrix} \tag{6}$$

l_1, l_2, l_4 are the length of the upper arm, forearm and palm. a_1, a_2, a_4 are the distances between the centroid of the upper arm, forearm and palm to their corresponding proximal joint. a_3, a_5 are the distances from the joint of the elastic absorber and viscous damper absorber to its corresponding proof mass. m_1, m_2, m_4 are the masses of the upper arm, forearm and the palm. m_{a_1}, m_{a_2} are the masses of viscous damper absorber and the elastic absorber of the SEVD absorber. I_1, I_2, I_4 are the mass moments of inertia of the upper arm, forearm and palm. M, C, K are the mass and the damping and stiffness coefficient matrices of the system. $\theta_1, \theta_2, \theta_3, \theta_{a_1}, \theta_{a_2}$ are the flexion angles at the shoulder, elbow and wrist joints.. k_1, k_2, k_3, k_4 and c_1, c_2, c_3, c_4 are the stiffness and damping coefficient of the shoulder, elbow, biceps brachii and wrist muscles. k_a, c_a are the stiffness and damping coefficients for beams of the elastic absorber and viscous damper absorber.

$$f = \{f_1 \quad f_2 \quad f_3 \quad 0 \quad 0\}^T \tag{7}$$

$$\begin{aligned} f_k &= F_{k_1} \cos(\omega_1 t) + F_{k_2} \cos(\omega_2 t), \; k = \{1,2,3\} \\ F_{k_1} &= F_{k_2} = 0.5 \, N.m \; and \; \omega_m = \omega_{n_m}, \; m = \{1,2\} \end{aligned} \tag{8}$$

f_1, f_2, f_3 are the input moments of the hand due to the shoulder, elbow and wrist joint muscles. F is the magnitude of the input moment and ω is the driving frequency

The natural frequencies obtained using primary system's characteristic equations are:

$$\omega_{n_1} = 3.609 Hz, \omega_{n_2} = 5.348 Hz \; and \; \omega_{n_3} = 12.682 Hz \tag{9}$$

III. TUNED ABSORBER'S DESIGN

Two tuned vibration absorbers (TVAs) are designed to be attached separately to the forearm at the same position. The two absorbers have the same total mass (251.2 *g*) and the same total length (8.5 *cm*).

As demonstrated by Gebai et al. [9], as the absorber's joint approaches the position of the wrist joint, more reduction can be achieved at the palm. Taking into consideration the maximum designed length of both absorbers, the absorbers will be tested at a position l_a which is 8.5 *cm* away from the wrist joint (l_a = l_2 – 8.5 *cm*). Absorbers dimensions can be chosen depending on the Dunkerley's semi-empirical formulation [18] as done by Gebai et al. [9][10][12].

The single DVA is modeled as a stainless steel alloy cantilevered beam with a copper mass attached along its length. Its configuration, dimensions and equivalent linear model are shown in Fig. 2. No damper is attached to this absorber, a very little damping can be provided by the beam's material. The stiffness (k_a) and damping (c_a) coefficients are assumed to be proportional by a constant [7], such that $c_a = 0.005 k_a$. This elastic absorber is tuned to the second natural (9) frequency of the primary system to reduce the resting tremor at this frequency:

$$\omega_a = \omega_2 \text{ and } \omega_2 = \omega_{n_2} \tag{10}$$

The dual SEVD vibration absorber is designed as an elastic absorber attached in series to a viscous damper absorber. The elastic absorber is formed from a thin cantilevered beam providing its stiffness (k_a) with a zero damping coefficient and a copper mass attached along its length. It is modeled at the tuning condition of (10). At the end of the elastic absorber's beam, a purely high damping material having a zero stiffness coefficient is modeled as beam's material. A copper mass is attached along the beam's end. The appropriate dimensions and equivalent linear model of this absorber are shown in Fig. 3. The damping coefficient (c_a) in the viscous damper absorber of the SEVD absorber is designed to satisfy the tuning condition of the fundamental frequency (9) of the primary system:

$$\omega_a = \omega_1 \text{ and } \omega_1 = \omega_{n_1} \tag{11}$$

The response of the system is needed for tuning the absorber at the chosen responses. The absorbers are tuned by satisfying the root in the real part (A_{1ik}) in the numerator of the corresponding response. The response is obtained using the transfer function (H) of this dynamically coupled system to represent the frequency domain response (Θ) using the Receptance transfer function (α) as follows:

$$H(\omega) = \{(-\omega^2[M] + [K]) + j\omega[C]\}^{-1} \tag{12}$$

$$H(\omega) = \{\alpha_1 \quad \alpha_2 \quad \ldots \quad \alpha_k\} \tag{13}$$

$$\alpha_k = \frac{A_{1_{ik}} + B_{1_{ik}}}{A_2 + jB_2} \tag{14}$$

$$\Theta_{ik} = \sum_{k=1}^{2} \sum_{m=1}^{2} |\alpha_k| F_{k_m} \tag{15}$$

i, k are the i-th row and k-th column for the $n \times n$ transfer function of the n-DOF system.

The single DVA and the elastic and viscous damper absorbers of the dual SEVD DVA are tuned to the wrist joint's response (Θ_3) due to the elbow muscle activation (F_2):

$$A_{1_{32}} = 0 \tag{16}$$

Using MATLAB, the stiffness and damping coefficients of each absorber can be determined as shown in Table 4, in addition to the absorbers mass.

All absorbers parameters are evaluated satisfying the tuning condition. Then, they are ready to be attached to the forearm to test their performance in tremor suppression.

IV. SIMULATED RESULTS

A. Frequency Domain

The frequency domain response of (15) is represented by graphs showing the behavior at the shoulder, elbow and wrist joints over a range of driving frequencies.

Maximum flexion angles are shown in Fig. 4 at the resonance frequencies of the uncontrolled hand system and the hand controlled by the single TVA and the dual series TVA. Tuning is well shown at the second natural frequency (tuning frequency) of the uncontrolled system at hand joints due to the single TVA. It shows high reduction in the tremor amplitude at the resonance frequencies of the proximal joint's responses and a good reduction at the other frequencies. The dual SEVD TVA shows qualitatively similar behavior to the single TVA with much damped tremor's amplitude. However, the wrist joint's response is subjected to tremor amplification at 7.623 Hz because of the high shifting to the right in the highest resonance frequency of the primary system due to its damper. The dual SEVD TVA is a very effective absorber at system's resonance frequencies and most of the other frequencies. It can cause a very high reduction in the resting tremor's amplitude in the hand of PD patients.

The resonance frequencies of the controlled system are derived using the characteristic equation, for the hand contr-

TABLE IV. TUNED ABSORBER'S PARAMETERS

Parameters	Single TVA	Dual series TVA	
		Elastic	Viscous Damper
$k_a\ (N.m/rd)$	0.2181	0.3211	0.0000
$c_a\ (N.m.s/rd)$	0.0011	0.0000	0.0016
$m_a\ (kg)$	251.2	125.6	125.6

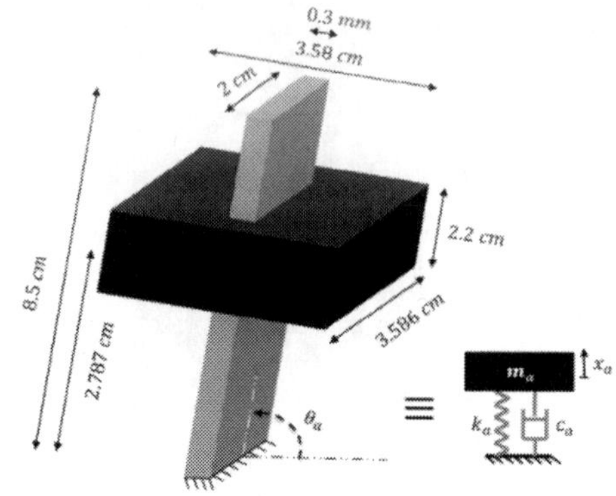

Figure 2. Single dynamic absorber.

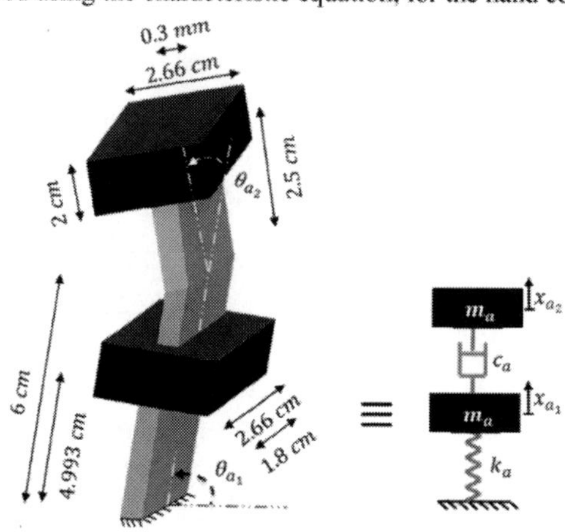

Figure 3. Dual series elastic-viscous damper absorber.

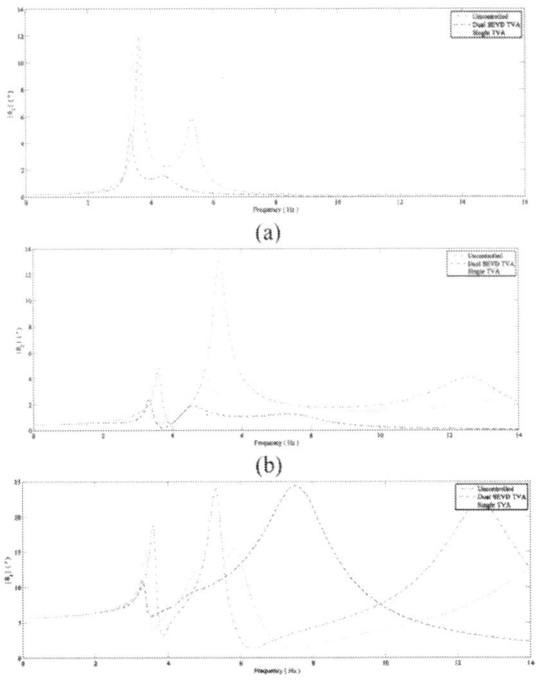

(a) (b) (c)

Figure 4. Frequency domain response at: shoulder (a), elbow (b) and wrist joints (c).

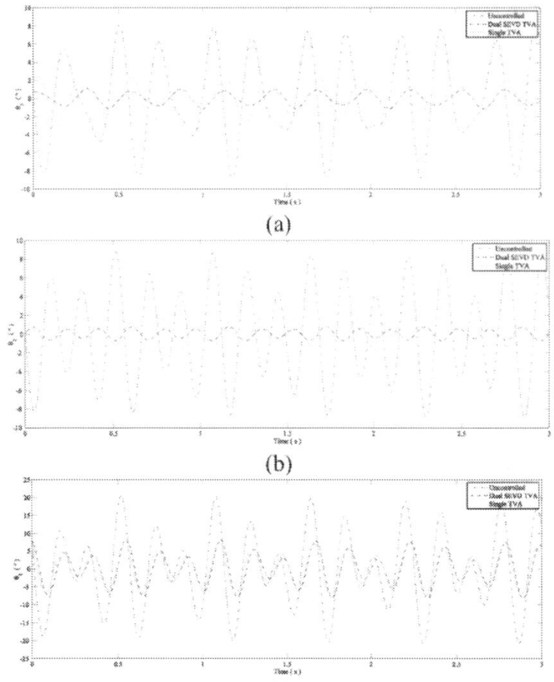

(a) (b) (c)

Figure 5. Time domain response at: shoulder, elbow and wrist joints.

olled by:

- Single TVA:

$$\omega_{n_1} = 3.263Hz, \omega_{n_2} = 3.921Hz, \omega_{n_3} = 5.383Hz, \quad and\ \omega_{n_4} = 13.915Hz \tag{17}$$

- Dual SEVD TVA:

$$\omega_{n_1} = 0Hz, \omega_{n_2} = 2.454Hz, \omega_{n_3} = 3.330Hz, \quad \omega_{n_4} = 4.496H\ and\ \omega_{n_5} = 7.623Hz \tag{18}$$

B. *Time Domain*

The time domain response is used to analyze the behavior of the system at the specified excitation frequencies in terms of joints angular displacement.

The time domain response is determined using (15) and derived as follows:

$$\theta_{ik} = \Theta_{ik} e^{j(\omega_m t - \varphi)}$$

$$|\Theta_{ik}| = F_{k_m} \sqrt{\frac{A_{1_{ik}}^2 + B_{1_{ik}}^2}{A_2^2 + B_2^2}} \quad and \quad \varphi = \tan^{-1}\left(\frac{B_{1_{ik}}}{A_{1_{ik}}}\right) \tag{19}$$

φ is the phase angle resulting from the damping coefficient. In Fig. 5a-c, the time domain response (19) at the shoulder, elbow and wrist joints responses of the hand excited at the first two resonance frequencies is shown due to the single joint muscles activation (8). The single TVA and dual SEVD TVA causes high reduction in tremor's flexion angle.

The percentage of reduction in the tremor's amplitude between the uncontrolled ($\Theta_{unc.}$) and controlled ($\Theta_{c.}$) systems in the time domain is calculated using this equation:

$$\%\,\text{Reduction} = \frac{\Theta_{unc.} - \Theta_{c.}}{\Theta_{unc.}} \times 100 \tag{20}$$

The percentage of reduction at the shoulder, elbow and wrist joints due to attaching both absorbers to the forearm is summarized in Table 5. It is shown that the dual SEVD TVA is an important absorber to be considered. It causes 85.7–86.6%, 88.4–91.3% and 58.29–59.9% reduction in the involuntary tremor's amplitude transmitted to the shoulder, elbow and wrist joints due to the activated muscles.

TABLE V. PERCENTAGE REDUCTION IN TREMOR'S AMPLITUDE

% Reduction	**Shoulder joint**	**Elbow joint**	**Wrist joint**
Single TVA	61.4–69.9	72.5–81.2	54.4–64.3
Dual TVA	85.7–86.6	88.4–91.3	58.29–59.9

The SEVD and single TVAs have the same total mass and length and are tested at the same design conditions. However, the DEVD dual TVA is more effective than the single TVA in reducing the flexion angular displacements.

V. DISCUSSION

Several studies [7][9]–[12][19] have been done to mechanically reduce the involuntary tremor in hands of PD patients. The main challenges are referred to the dynamic hand modeling that can best reflect Parkinsonism and that takes into account the tremor displacement at most allowed angular displacements. In addition, developments in the absorbers design are of high importance. Changing in the configuration of the passive absorber aiming to increase its performance attracted many researchers [8][12].

In a previous study, Hashemi *et al.* [7] have modeled the human hand as two uniform rigid rods to describe the flexion-extension motion of the elbow and shoulder joints. The system was excited due to the elbow muscle activation driven at a single frequency. In our study, the hand system is modified considering an additional DOF to describe the angular displacement at the wrist joint of a system actuated with several muscles. In [7], a one-DOF absorber was attached to the forearm but it was not designed at the tuning condition. The resonance frequency was 2.24 *Hz* while the absorber's natural frequency was 2.755 *Hz*. So, the absorber was not designed to operate at its maximum performance.

Rahnavard et al. [19] used the same hand model designed by Hashemi et al. [7]. However, the single DOF absorber's parameters were designed using the H_2 optimization method. The proposed system leads to a 20 *cm* long absorber providing high percentage of reduction at the shoulder and elbow joints. In our study, very effective single and dual TVAs are designed with an optimum length of 8.5 *cm* at all the hand joints.

VI. Conclusion

The three-DOF dynamic model of the human hand is designed to reflect the biodynamic response at the proximal joints of an elderly PD patient. The model describes the flexion-extension motion for the system operating at resting tremor's resonance frequencies in the horizontal plane. Two absorbers having the same total mass and length are tuned to reduce the involuntary tremor transmitted to the hand joints due to the single joint muscles activation. Both absorbers are tested when attached at the same position on the forearm. The one-DOF absorber causes 61.4–69.9%, 72.5–81.2% and 54.4–64.3% reduction in tremor amplitude at the shoulder, elbow and wrist joints in the time domain. A very effective two-DOF SEVD TVA is used for tremor reduction. The SEVD is formed from a series combination between the elastic absorber and the viscous damper absorber. It causes 85.7–86.6%, 88.4–91.3% and 58.29–59.9% reduction at the shoulder, elbow and wrist joints.

As future work, the molded hand system can be sized depending on the hand dimensions of a real Parkinson patient. Then, the suggested passive controllers can be manufactured and tested when attached to the forearm of the real hand. In addition, a comparison can be done between the numerical and experimental studies.

References

[1] S. Morrison, G. Kerr, and P. Silburn, "Bilateral tremor relations in Parkinson's disease: Effects of mechanical coupling and medication," *Parkinsonism & related disorders,* vol. 14, no. 4, pp. 298-308, 2008.

[2] S. K. Van Den Eeden, C. M. Tanner, A. L. Bernstein, R. D. Fross, A. Leimpeter, D. A. Bloch and L. M. Nelson, "Incidence of Parkinson's disease: variation by age, gender, and race/ethnicity," *American journal of epidemiology,* vol. 157, no. 11, pp. 1015-1022, 2003.

[3] E. Rocon, J. M. Belda-Lois, and J. J. Sanchez-Lacuesta, "Pathological tremor management: Modelling, compensatory technology and evaluation," *Technology and Disability,* vol. 16, no. 1, pp. 3-18, 2004.

[4] G. M. Hariz and L. Forsgren, "Activities of daily living and quality of life in persons with newly diagnosed Parkinson's disease according to subtype of disease, and in comparison to healthy controls," *Acta Neurologica Scandinavica,* vol. 123, no. 1, pp. 20-27, 2011.

[5] K. M. Jackson, J. T. Joseph, and S. J. Wyard, "A mathematical model of arm swing during human locomotion," *Journal of biomechanics,* vol. 11, no. 6-7, pp. 277-289, 1978.

[6] R. Raikova, "A general approach for modelling and mathematical investigation of the human upper limb," *Journal of biomechanics,* vol. 25, no. 8, pp. 857-867, 1992.

[7] S. M. Hashemi, M. F. Golnaraghi, and A. E. Patla, "Tuned vibration absorber for suppression of rest tremor in Parkinson's disease," *Medical and Biological Engineering and Computing,* vol. 43, no. 1, pp. 61-70, 2004.

[8] T. Igusa and K. Xu, "Vibration control using multiple tuned mass dampers," *Journal of sound and vibration,* vol. 175, no. 4, pp. 491-503, 1994.

[9] S. Gebai, M. Hammoud, A. Hallal, A. AL Shaer, and H. Khachfe, "Passive Vibration Absorber for Tremor in Hand of Parkinsonian Patients," in *3rd International Conference on Advances in Computational Tools for Engineering Applications, ACTEA2016,* 2016.

[10] S. Gebai, M. Hammoud, and H. Khachfe, "Mechanical Vibration Absorber to Reduce the Neurodegenerative Disorder of Elderly," in *4th International Conference on Advanced Training and Lifelong Learning Programs in Applied Health Sciences: Elderly Care,* 2016.

[11] S. Gebai, M. Hammoud, A. Hallal, A. Shaito, and A. AL Shaer, "A Dual Dynamic Vibration Absorber to Attenuate Rest Tremor at the Hand of a Parkinson's Patient," in *22nd International Scientific Conference of the Social Avenues of Research, LAAS,* 2016.

[12] S. Gebai, M. Hammoud, A. Hallal, and H. Khachfe, "Tremor Reduction at the Palm of a Parkinson's Patient Using Dynamic Vibration Absorber," *Bioengineering,* vol. 3, no. 3, pp. 18, 2016.

[13] F. E. Zajac and M. E. Gordon, "Determining muscle's force and action in multi-articular movement," *Exercise and sport sciences reviews,* vol. 17, no. 1, pp. 187-230, 1989.

[14] R. Drillis, R. Contini, and M. Bluestein, "Body segment parameters," *Artificial limbs,* vol. 8, no. 1, pp. 44-66, 1964.

[15] N. M. M. Maia and J. M. e Silva, "Theoretical and experimental modal analysis," 1997.

[16] V. Wright and R. J. Johns, "Physical factors concerned with the stiffness of normal and diseased joints," *Bulletin of the Johns Hopkins Hospital,* vol. 106, p. 215, 1960.

[17] G. Deuschl, J. Raethjen, M. Lindemann, and P. Krack, "The pathophysiology of tremor," *Muscle & nerve,* vol. 24, no. 6, pp. 716-735, 2001.

[18] H. H. Jeffcott, "The Periods of Lateral Vibration of Loaded Shafts. The Rational Derivation of Dunkerley's Empirical Rule for Determining Whirling Speeds," *Proceedings of the Royal Society of London. Series A, Containing Papers of a Mathematical and Physical Character,* vol. 95, no. 666, pp. 106-115, 1918.

[19] M. Rahnavard, M. Hashemi, F. Farahmand, and A. F. Diza, "Designing a hand rest tremor dynamic vibration absorber using H 2 optimization method," *Journal of Mechanical Science and Technology,* vol. 28, no. 5, pp. 1609-1614, 2014.

Towards Evidence-Based Self-Management for Spondyloarthritis Patients

Wolfgang Leister, Ingvar Tjøstheim,
Trenton Schulz, Aleksander Bai
Norsk Regnesentral
Oslo, Norway
email: {wolfgang.leister, ingvar.tjostheim,
trenton.schulz, aleksander.bai}@nr.no

Kåre Birger Hagen, Camilla Fongen,
Nina Østerås, Hanne Dagfinrud
NKRR, Diakonhjemmet Hospital
Oslo, Norway
email: k.b.hagen@medisin.uio.no
camilla.fongen@diakonsyk.no

Lars Thomas Boye
Tellu AS
Asker, Norway

Vebjørn Berre
APERTUS AS
Asker, Norway

Liming Chen
De Montfort University
Leicester, UK
email: liming.chen@dmu.ac.uk

Dag Magnar Soldal, Carl-Henrik Franke
Hospital of Southern Norway
Kristiansand, Norway
email: {dag.magnar.soldal,
carl-henrik.franke}@sshf.no

Jon Hagfors,
Jan Halvard Relbe-Moe
Norsk Revmatikerforbund
Norway

***Abstract*—We developed a concept including a set of tools for self-management for patients suffering from axial spondyloarthritis (SpA). This concept involves patient-recorded outcome measures, both subjective assessment and clinical measurements, that are used to present recommendations. We report from experiences made while implementing a proof of this concept. Besides giving the patient a self-management tool, our work also improved the methodology for clinical measurements.**

Keywords*—*axial spondyloarthritis; self-mangement; healthcare; self-assessment; evidence-based; mobile applications.

I. Introduction

For a variety of chronic diseases, patients managing the condition themselves (self-management) can result in reduced costs in the health care sector and an improved clinical outcome [1]. Self-management encompasses methods where the patient participates in managing their disease through education and changes of behaviour and lifestyle [2] [3]. In evidence-based self-management, elements such as clinical assessment, collaborative priority and goal-setting, patient self-efficacy, and active follow-up are essential [4]. We look closer at self-management settings [5] where patients assess the status of their disease using sensors and questionnaires on their smartphones and report the results (i.e., patient-reported outcome measures) [6] [7]. Based on the results patients receive non-pharmacological recommendations from the self-management system to increase their coping skills, help with pain management, adhere to their medication regime, improve self-care behaviours, and enact lifestyle changes.

Spondyloarthritis (SpA) describes a group of several related, but phenotypically distinct rheumatic diseases, such as ankylosing spondylitis (AS). The condition axial SpA is characterised by inflammatory back pain and mainly affects the axial skeleton, which is distinct from peripheral SpA with other symptoms. In axial SpA, the first appearance is mainly in young adulthood and can lead to structural and functional impairments and a decrease in health related quality of life. Although axial SpA is a chronic condition, the symptoms and disease activity vary over time [8]. The primary goals for managing axial SpA are to maximise long term health-related quality of life by controlling symptoms and inflammation, preventing progressive structural damage in the spine, and normalising function and social participation. Relevant medication and physical training are recommended as the foundation of the management of axial SpA [9].

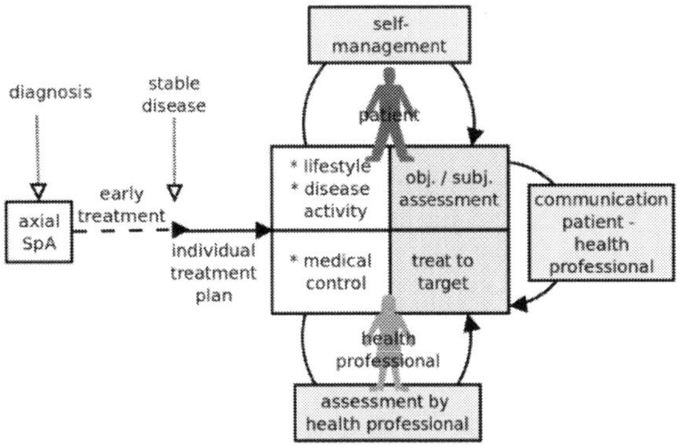

Fig. 1: Concept for controlling the disease with the three parts: self-management, clinical assessment, and patient-health personnel communication.

Currently, there are few self-management tools for axial SpA that are evidence-based. Some tools for subjective assessment exist, but sensor-based tools for objective assessment are not yet available to the wider public. Also, there are obstacles to let patient-assessed data be of use in a clinical setting [10].

This paper presents a concept for evidence-based, self-management of axial SpA, supported by an implementation of a smartphone and sensor-based system that can give recommendations to the patients. We report from experiences from this implementation.

The remainder of this paper is organised as follows: After a brief presentation of related work (Section II) and presenting our concept of self-management for axial SpA patients (Section III), we show details from the proof of concept implementation involving subjective and sensor-based clinical assessment and recommendations to the patient (Section IV). Further, we present the results from a usability test (Section V). We discuss our findings (Section VI) and conclude in Section VII.

II. RELATED WORK

Considerable work has been done on self-management programs for chronic diseases with good results in terms of quality of life, and reducing the need for care and cost efficiency [11]. Programs such as *The Chronic Disease Self-Management Program* have shown significant improvement in health distress and increased perceived self-efficacy [12]. The motivation for these programs is to provide people with chronic diseases the tool to efficiently mange their own condition.

We can find management support for some chronic conditions using information and communication technology (ICT). These include: a self-management application called SoberDiary for alcoholism [13], a mobile application for diabetes that integrates with personal smartwatches [14], a virtual coach for chronically ill elderly [15], a smartphone app for rheumatic diseases management [16]. There are also generic apps for integration vital signs into personal health devices or electronic medical record systems [17].

Within axial SpA there are ICT apps like SpA Helper [18] that helps monitor the disease. But the results from the monitoring are not part of a feedback cycle; i.e., they do not use the treat-to-target principle (Section III-A).

III. SELF-MANAGEMENT

We find multiple definitions of *self-management* in the literature. But we prefer the definition by Barlow et al. [19] "...the individual's ability to manage the symptoms, treatment, physical and psychosocial consequences and life style changes inherent in living with a chronic condition." Barlow et al. stress that monitoring one's condition and the effect of responses to daily life can lead to a dynamic and continuous process of self-regulation.

A. Integrating Treat-To-Target

The *treat-to-target method* [20] has been developed for treating axial SpA. This evidence-based method is used after diagnosis and early treatment, when the disease has reached a stable state (Fig. 1). At this stage, an individual treatment plan has been created for the patient. This method uses a treatment goal (*target*) for a *treatment plan*. Following the treatment plan and regularly assessing the patient's status provides evidence about how the disease develops. When the patient's status moves away from the treatment target to a worse condition, health personnel, in discussion with the patient, might adjust the treatment plan or target.

As part of an evidence-based, self-management setting, the treat-to-target method is extended so the patient can perform self-assessment to gather evidence about the current disease condition by performing assessments, answering questionnaires,

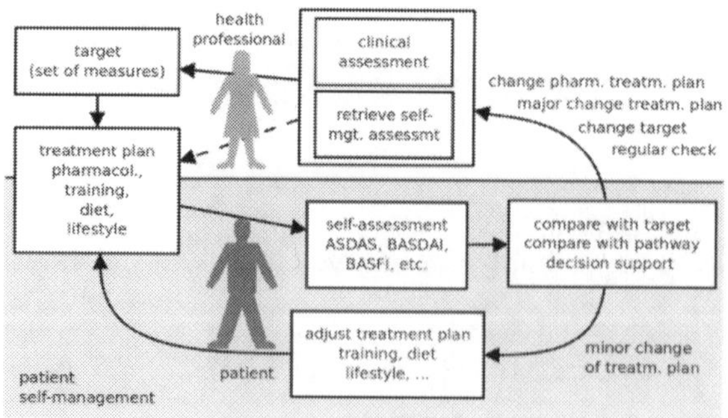

Fig. 2: Treat-to-target in a self-management setting, showing tasks to be performed by patient and health personnel, respectively.

and following the progress from the patient diary. The patient diary data can be used for patient-health personnel communication by making it available to the clinical personnel, either regularly or when needed (e.g., a patient visit).

Fig. 2 shows how treat-to-target can be aligned with self-management. The upper unshaded part of the drawing is the health personnel domain. This is where health personnel perform clinical assessments and decide the treatment target and treatment plan. The lower shaded part is the patient's domain. This is where the patient can perform assessments, compare with the target, and adjust some elements of the treatment, e.g., physiotherapy or exercise.

B. An Architecture for axial SpA Treatment

Our concept (Fig. 3) builds on *a*) a solution for self-management, *b*) better quality and effectiveness of *clinical assessment*, and *c*) enhanced *patient-health personnel communication*.

The solution for self-management lets the patients use tools at home to control the disease. It includes patient-reported outcome measures [7], the assessment of ample parameters, the use of a patient diary [21], patient guidance with respect to the treat-to-target principles, and alerts in case of changes of the patient's condition or physical function.

The concept also enhances the quality and effectiveness of clinical assessment; assessment methods developed for self-management are made available for clinical assessment.

The concept includes a foundation for patient-health personnel communication. Self-reported assessments can be used for patient-health personnel communication to explain or visualise the development of the disease and data transfer to the hospital.

IV. PROOF OF CONCEPT FOR AXIAL SPA SELF-MANAGEMENT

The parts of this architecture that include data exchange between a health cloud or a patient's devices and the electronic health record (EHR) system are beyond the scope of our work. These are parts that rely on policies defined by public healthcare providers. So, we focused on implementing tools for clinical assessment and self-management.

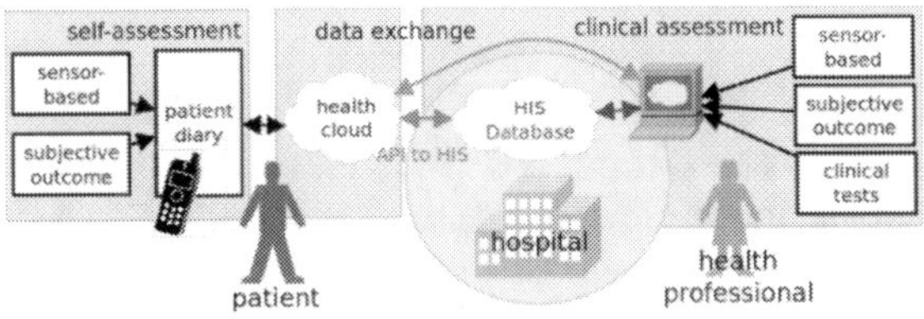

Fig. 3: Architecture of a self-management system including three parts: self-management, clinical assessment, and data exchange.

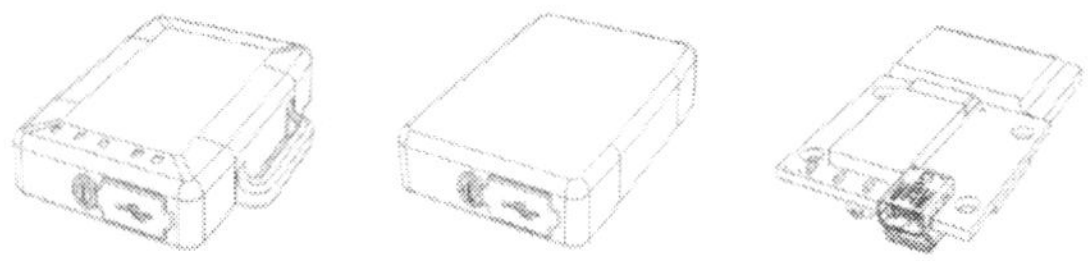

Fig. 4: Drawing of the APERTUS sensor used for axial movements.

A. Medical Assessment Methods for axial SpA

Medical self-assessment is essential for evidence-based self-management. So, these self-assessment methods should be based on medical assessment methods since evidence for their effectiveness is documented.

The *AS Disease Activity Score* (ASDAS) is used for measuring and monitoring disease activity in axial SpA. It is based on a composite score of domains relevant to patients and clinicians, including both self-reported items and objective measures [22].

The Bath indices [23] present outcome measures for use with SpA patients, and consist of four indices: the Bath AS Metrology Index (BASMI), the Bath AS Functional Index (BASFI), the Bath AS Disease Activity Index (BASDAI), and the Bath AS Patient Global Score (BAS-G). These indices are designed to give a good clinical assessment using a minimum number of measurements or questions to be answered. The BASMI is five simple clinical measurements; the other indices consist of a number of questions that are answered on a rating scale from zero to ten.

Østerås et al. [24] described a set of assessment tests that are candidates for axial SpA self-assessment. These exercises include: lateral spinal flexion, modified Schober's, cervical rotation, occiput to wall distance, tragus to wall distance, intermalleolar rotation around the vertical axis, lumbal/thorcal rotation, six-minutes walking test, stair climb test, sit-to-stand test, fingertip-to-floor test, and maximum grip strength test.

B. Sensor-based clinical measurements

APERTUS developed a sensor that can measure rotation around the vertical axis such as cervical rotation, thoracic rotation and hip abduction (measured in the supine position). Furthermore, the result can be transmitted via a wireless connection to a receiver, such as PC, tablet, or smartphone. This inertial sensor is packaged in a small box (Fig. 4) that can be attached to the body. The size of the device is $55\text{mm} \times 35\text{mm} \times 3\text{mm}$. The sensor contains radio technology that follows Bluetooth standards that might influence electronic devices in 2.4 GHz ISM, but to a significantly lower degree than mobile phones.

Compared with other technology such as lasers or optical sensors, this sensor's advantages include its high precision and being cheaper, smaller, and lighter than the other solutions. Compared to the traditional way of measuring rotation with a goniometer or myrinometer (e.g., compass) the sensor provides more precise measurements. The sensor is a simple way to achieve satisfactory measurements in acceptable use of time and without health personnel assisting.

In a laboratory setting, the sensor has excellent criterion validity and reliability for rotation around the vertical axis in the range of motion from 10 to 120 degrees. The angle can be measured with a precision of $\pm 1.3°$. These findings justify proceeding with further evaluations of the sensor for this kind of measurements [25] [26]. A clinical trial of the rotation measurements with 60 patients suffering from axial SpA is currently under evaluation.

We developed a suitable user interface for the assessment process with the sensor. The assessed data are stored locally in the patient diary and forwarded to the health cloud for permanent storage.

C. Self-Assessment of Subjective Conditions

For the assessment of the subjective conditions for BASDAI and BASFI we implemented suitable user interfaces in our prototype (Fig. 5). We also implemented a questionnaire for ASDAS, a composite score including subjective evaluation and the inflammatory markers C-reactive protein (CRP) and Erythrocyte sedimentation rate (ESR). Patients answer The questions on a scale from zero to ten by tapping on the appropriate number. We did not choose sliders because we assumed that tapping on the appropriate field would be easier for the target group with their possible movement restrictions.

After the form is finished, the data are stored locally. An estimate of the current health condition is shown to give the patient feedback along with the possibility to report these data to the health cloud for permanent storage.

Questionnaires can be scheduled using the mobile device's calendar by creating calender entries with a specific syntax. The calendar then reminds patients to perform assessments at a given time.

D. Self-Management and Recommendation

A self-management system needs to support the patient in the following ways: *a*) deciding the type and degree of adjustments for non-pharmacological changes in a treatment plan, such as diet, training, lifestyle, or other minor adjustments; *b*) identifying significant deviations from the expected progress and present these deviations to the patient and health personnel; *c*) advising changes of treatment plan to the health personnel; and *d*) suggest changes of patient's target to the health personnel. Qi et al. [27] present an approach for how to make decisions that are presented to the patient. We use a diary in our solution.

(a) The ASDAS/BASDAI menu. (b) The BASFI menu. (c) The result page.

Fig. 5: Screenshots for the data collection module.

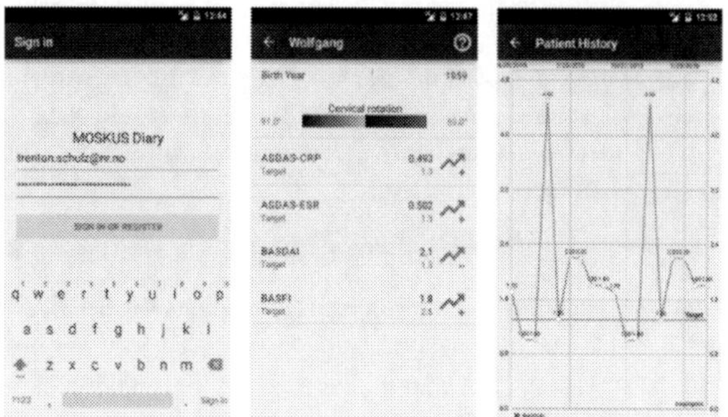

(a) The diary login screen. (b) The diary patient view. (c) The diary history view for BASDAI.

Fig. 6: Screenshots from the patient diary app.

The diary shows the disease's development visually, deviations from the treatment plan, and gives recommendations using *trend labels*. The patient view (Fig. 6b) shows the patient's birth year, the current left and right cervical rotation, and the current scores for ASDAS, BASDAI, and BASFI, including their targets. Each of these scores can have historical data; this is shown in the patient history view (Fig. 6c) and summarized in the trend labels (Fig. 7) to the right of the value. There are five trend labels: (*a*) disease activity is increasing, but below target; (*b*) disease activity is increasing; (*c*) no change in disease activity; (*d*) disease activity is decreasing, either below or heading towards target; and (*e*) disease activity is decreasing, but still very high and needs more treatment.

(a) (b) (c) (d) (e)

Fig. 7: Trend labels for the patient view.

E. Heuristic Decision Support Based on Medical Expertise

We created simple rules to guide patients and health personnel. These rules are based on the values from ASDAS, BASDAI, and BASFI.

The values derived in ASDAS and BASDAI indicate the amount of disease activity. Machado et al. [28] define cutoff values for disease activity measured using ASDAS: *a*) under 1.3 the disease is inactive; *b*) between 1.3 and 2.1 disease activity is moderate; *c*) 2.1 to 3.5 disease activity is high; and *d*) Over 3.5 disease activity is very high. A change on the ASDAS scale of 1.1 or more is considered a clinically important change while 2.0 or more is considered a major change. Based on this work, we indicate the trend of the scores (up, down, or steady), as well as the severity (colour). The thresholds can be personalised for patients where health personnel defines alternative values.

Braun et al. [29] propose a similar approach for BASDAI by calculating a trend line that uses BASDAI targets for cutoffs. Situations where the BASDAI is above 4.0 - indication of high disease activity - or changes of the BASDAI over 50% or a factor of two also generate a warning to contact health personnel.

BASFI indicates the disability level. Wariaghli et al. [30] ran a large survey with Moroccan patients and defined the target values depending on the patient's age in three age ranges. We use similar rules as above for determining the trend based on the patient's target or age information, depending on what is available.

V. Usability Test of the Prototypes

Usability, how easy something is to use, is an important factor for adoption and continuous use of a system or application. Motivated by this, we performed a usability test of the two developed prototypes. We wanted to see how appropriate the apps are for their purpose, and to get feedback on the usability. We employed the System Usability Scale (SUS) developed by Brooke [31]. The SUS consists of ten questions that are rated by the participants on a five point or a seven point Likert scale [32]. The ratings are used to calculate a score on a scale from 0 to 100 where 70 is the average score. Additionally, we added six questions on related matters that are not part of the SUS scale, e.g., the need for the apps, satisfaction, and whether participants would recommend the apps to others.

For the usability test, we recruited eighteen individuals with Android smartphones among members of the Norwegian Rheumatology Association (Norsk Revmatikerforbund). We asked the eighteen to download the two apps and sign up for the usability test. Of the invited participants, fourteen followed the procedure, downloaded the app, and registered at the health cloud site. The individuals received the link to the surveys after they had downloaded the apps and a text message with instructions. Of the eighteen individuals, nine used the apps and completed the test.

With only nine respondents, the usability test is more a pilot study. If an application has major usability weaknesses, these will likely be revealed with small sample sizes also. Our

TABLE I: SUS scores for the collection module app and patient diary app with highest and lowest scores removed.

	Average	Median
Collection module	73.73	80.20
Patient diary	74.05	74.35

test did not indicate major weaknesses. We did not perform statistical analysis of the data beyond calculating SUS scores.

We performed separate tests for each app. The average and median from the SUS are presented in Table I with the highest and lowest scores eliminated. The SUS scores of the apps are 73 and 74, respectively.

VI. Discussion

The proposed concept for self-management is based on a feedback loop with the patient is involved. Axial SpA does not require immediate attention when the condition worsens, but an appointment with a clinic needs to be scheduled. Also, not adhering to the self-management regime does not have other side-effects beyond not adhering to the treatment, and these patients need to keep the conventional frequency of clinical follow-ups. Note that other chronic diseases might require immediate attention in some situations or not adhering to the self-management regime might worsen the patient's condition. Thus, an evaluation is needed for other conditions than axial SpA to see if our self-management architecture can be applied.

Data assessed in self-management are usually not complete or might be of a different nature in terms of the clinical indices. For example, the values extracted from blood samples might not be available, only selected values from the BASMI examinations might be available, or the patient assesses alternative measurements that are not part of the established indices. To support recommendations in these cases, it is necessary to predict an individuals axial SpA disease condition based on a combination of physiological, behavioural and subjective (self-reported) features. To achieve this, Schiboni et al. [33] have proposed a fuzzy rule-based evidential reasoning (FURBER) approach for multiple assessment fusion. But this approach requires enough real patient data as training data to be considered for real treatment.

The medical indices for axial SpA and the data retrieved from the FURBER method are only suited to give an indication of the disease conditions at one moment. For predicting the probable development of the patient's health condition and whether actions need to be taken requires temporal reasoning. Modelling the disease development as a stochastic process to optimise the treatment recommendations could be done by a Markov Decision Process (MDP) [34]. Yet a large sample size could make this approach less viable [35]. Alternatively, the *patient profiling* method described by Lutz et al. [36], could be feasible.

The new assessment methodology for rotation exercises using sensor technology will also impact clinical use as it will save time and provide better results. Today, health personnel use goniometer or compass-based measurements that are time consuming but have acceptable accuracy. The trials in clinics have shown this new methodology simplifies clinical measurements, greatly improves accuracy, and saves time for the health personnel and the patient. The time saved and higher-quality data quickly make up for the cost of the sensors. Specifically, the much higher accuracy and easier handling of the sensor technology compared to the traditionally used methods is attractive to health professionals. Furthermore, the sensor will enable patients to perform the measurements themselves without the involvement of health personnel.

VII. Conclusion

We presented an architecture for self-management of axial SpA patients that is based on self-assessment by these patients. We have performed a proof of concept by implementing vital parts of a self-management system including clinical measurements, patient-reported outcome measurements, feedback module, patient diary, and decision making software.

Further user evaluations will be necessary before a system based on our architecture can be brought into clinical practise. In addition, communication modules to the EHR system of the clinics need to be implemented. Further, the development of suitable measurements for exercises beyond rotation exercises need to be developed in a way that allows patients to perform these at home.

Finally, since patient-reported data might not be of the best quality (e.g., they have not undergone quality assurance or might be incomplete) estimation methods both for the current disease status and for temporal prediction need to be developed. While we could show the viability of the methods, further implementation work needs to be done.

VIII. Acknowledgments

The work presented here has been carried out in the **M**obile muscul**O** **SK**eletal **U**ser **S**elf-management (MOSKUS) project funded by the Research Council of Norway in the VERDIKT programme, grant number 227251.

References

[1] L. C. Li, A. F. Townsend, and E. M. Badley, "Self-management interventions in the digital age: New approaches to support people with rheumatologic conditions," Best Pract Res Clin Rheumatol, vol. 26, no. 3, pp. 321–33, 2012.

[2] K. R. Lorig, P. D. Mazonson, and H. R. Holman, "Evidence suggesting that health education for self-management in patients with chronic arthritis has sustained health benefits while reducing health care costs," Arthritis & Rheumatism, vol. 36, no. 4, pp. 439–446, 1993.

[3] T. J. Brady, J. Kruger, C. G. Helmick, L. F. Callahan, and M. L. Boutaugh, "Intervention programs for arthritis and other rheumatic diseases," Health Education & Behaviour, vol. 30, no. 1, pp. 44–63, February 2003.

[4] M. Battersby et al., "Twelve Evidence-Based Principles for Implementing Self-Management Support in Primary Care," Joint Commission Journal on Quality and Patient Safety, pp. 561–570, Dec. 2010.

[5] Y. Miedany, "e-rheumatology: are we ready?" Clinical Rheumatology, vol. 34, no. 5, pp. 831–837, 2015.

[6] D. Palmer and Y. E. Miedany, "Proms: a novel approach to arthritis self-management," British Journal of Nursing, vol. 21, no. 10, pp. 601–607, 2012.
[7] Ø. Dale, T. Schulz, A. Christie, and H. Dagfinrud, "Collection of patient reported outcome measures using short messaging service," in eTELEMED 2013, The Fifth International Conference on eHealth, Telemedicine, and Social Medicine, pp. 214–219. International Academy, Research and Industry Association (IARIA), 2013.
[8] J. Braun and J. Sieper, "Ankylosing spondulitis," Lancet, vol. 369, pp. 1379–1390, 2007.
[9] J. Braun et al., "2010 update of the ASAS/EULAR recommendations for the management of ankylosing spondylitis," Annals of the Rheumatic Diseases, vol. 70, no. 6, pp. 896–904, 2011.
[10] P. West, R. Giordano, M. van Kleek, and N. Shadbolt, "The Quantified Patient in the Doctor's Office," in Proceedings of the 2016 CHI Conference on Human Factors in Computing Systems - CHI '16, pp. 3066–3078. New York, USA: ACM Press, 2016.
[11] U. E. Bauer, P. A. Briss, R. A. Goodman, and B. A. Bowman, "Prevention of chronic disease in the 21st century: elimination of the leading preventable causes of premature death and disability in the usa," The Lancet, vol. 384, no. 9937, pp. 45–52, 2014.
[12] K. R. Lorig et al., "Chronic disease self-management program: 2-year health status and health care utilization outcomes," Medical care, vol. 39, no. 11, pp. 1217–1223, 2001.
[13] C. You et al., "Soberdiary: A phone-based support system for assisting recovery from alcohol dependence," in Proceedings of the 33rd Annual ACM Conference on Human Factors in Computing Systems, CHI 2015, Seoul, Republic of Korea, April 18-23, 2015, B. Begole, J. Kim, K. Inkpen, and W. Woo, Eds., pp. 3839–3848. ACM, 2015.
[14] E. Årsand, M. Muzny, M. Bradway, J. Muzik, and G. Hartvigsen, "Performance of the First Combined Smartwatch and Smartphone Diabetes Diary Application Study," Journal of Diabetes Science and Technology, vol. 9, no. 3, pp. 556–563, May 2015.
[15] A. van Wissen, C. Vinkers, and A. van Halteren, "Developing a virtual coach for chronic patients: A user study on the impact of similarity, familiarity and realism," in Persuasive Technology, pp. 263–275. Springer, 2016.
[16] A. R. P. Azevedo, H. M. L. de Sousa, J. A. F. Monteiro, and A. R. N. P. Lima, "Future perspectives of smartphone applications for rheumatic diseases self-management," Rheumatology international, vol. 35, no. 3, pp. 419–431, 2015.
[17] H. S. Park, H. Cho, and H. S. Kim, "Development of a multi-agent m-health application based on various protocols for chronic disease self-management," Journal of medical systems, vol. 40, no. 1, pp. 1–14, 2016.
[18] M. Jagoda, "SpA Helper," web pages. [Online]. Available: https://itunes.apple.com/cv/app/spa-helper/id722932133 [Accessed: 2. May, 2016].
[19] J. Barlow, C. Wright, J. Sheasby, A. Turner, and J. Hainsworth, "Self-management approaches for people with chronic conditions: a review," Patient Education and Counseling, vol. 48, no. 2, pp. 177–187, 2002.
[20] J. S. Smolen et al., "Treating spondyloarthritis, including ankylosing spondylitis and psoriatic arthritis, to target: recommendations of an international task force." Ann Rheum Dis, vol. 73, no. 1, pp. 6–16, 2014.
[21] S. A. Raymond and R. N. Ross, "Electronic subject diaries in clinical trials," Applied Clinical Trials, pp. 48–58, March 2000.
[22] J. Zochling, "Measures of symptoms and disease status in ankylosing spondylitis: Ankylosing Spondylitis Disease Activity Score (ASDAS), Ankylosing Spondylitis Quality of Life Scale (ASQoL), Bath Ankylosing Spondylitis Disease Activity Index (BASDAI), Bath Ankylosing Spondylitis Functional Index (BASFI), Bath Ankylosing Spondylitis Global Score (BAS-G), Bath Ankylosing Spondylitis Metrology Index (BASMI), Dougados Functional Index (DFI), and Health Assessment Questionnaire for the Spondylarthropathies (HAQ-S)," Arthritis Care Res.(Hoboken.), vol. 63 Suppl 11, pp. S47–S58, Nov 2011.
[23] K. Irons and C. Joffries, "The Bath Indices – outcome measures for use with ankylosing spondylitis patients," The National Ankylosing Spondylitis Society (NASS), Surrey, UK, 2009.
[24] N. Østerås, H. Dagfinrud, A. T. Tveter, D. M. Soldal, C. Fongen, and C.-H. Franke, "About spondyloarthritis, disease specific questionnaires and assessments, and generic field tests for health-related physical fitness," Diakonhjemmet Sykehus, NKRR-notat 2-2013, 2013.
[25] C. Fongen, A. T. Tveter, F. R. Andersen, N. Østerås, and V. Berre, "Laboratory validity and reliability of a new sensor developed for measuring rotation around the vertical axis – a subproject in the Mobile mosculO SKeletal User Self-management," Diakonhjemmet Sykehus, NKRR-notat 1-2016, 2016.
[26] C. Fongen, N. Østerås, F. R. Andersen, V. Berre, A. T. Tveter, and K. B. Hagen, "Innovation in health care – laboratory validity and reliability of a new sensor developed to measure cervical rotation," Annals of the Rheumatic Diseases, vol. 75, no. suppl 2, p. 1288, 2016.
[27] J. Qi, L. Chen, W. Leister, and S. Yang, "Towards knowledge driven decision support for personalized home-based self-management of chronic disease," in Proc. of the 2015 Smart World Congress, pp. 1724–1729, 2015.
[28] P. Machado et al., "Ankylosing spondylitis disease activity score (asdas): defining cut-off values for disease activity states and improvement scores," Annals of the rheumatic diseases, vol. 70, no. 1, pp. 47–53, 2011.
[29] J. Braun et al., "International ASAS consensus statement for the use of anti-tumour necrosis factor agents in patients with ankylosing spondylitis." Annals of the rheumatic diseases, vol. 62, no. 9, pp. 817–824, 2003.
[30] G. Wariaghli et al., "Normative values for the Bath Ankylosing Spondylitis Functional Index in the general population compared with ankylosing spondylitis patients in Morocco," BMC Musculoskelet Disord, vol. 13, no. 1, p. 40, 2012.
[31] J. Brooke, "SUS: A quick and dirty usability scale," in Usability evaluation in industry, P. W. Jordan, B. Weerdmeester, A. Thomas, and I. L. McClelland, Eds. London: Taylor and Francis, 1996.
[32] K. Finstad, "Response interpolation and scale sensitivity: Evidence against 5-point scales," J. Usability Studies, vol. 5, no. 3, pp. 104–110, May 2010.
[33] G. Schiboni, W. Leister, and L. Chen, "Using fuzzy evidential reasoning for multiple assessment fusion in spondylarthropathic patient self-management," in Emerging Trends and Advanced Technologies for Science and Information, ser. Studies in Computational Intelligence. Springer, 2016.
[34] C. C. Bennett and K. Hauser, "Artificial intelligence framework for simulating clinical decision-making: A markov decision process approach," Artificial Intelligence in Medicine, vol. 57, no. 1, pp. 9–19, 2013.
[35] A. J. Schaefer, M. D. Bailey, S. M. Shechter, and M. S. Roberts, "Modeling medical treatment using markov decision processes," in Operations Research and Health Care: A Handbook of Methods and Applications, M. L. Brandeau, F. Sainfort, and W. P. Pierskalla, Eds., pp. 593–612. Boston, MA: Springer US, 2004.
[36] W. Lutz, Z. Martinovich, and K. I. Howard, "Patient profiling: An application of random coefficient regression models to depicting the response of a patient to outpatient psychotherapy." Journal of consulting and clinical psychology, vol. 67, no. 4, p. 571, 1999.

Aging Measurements with Precise Observations of Synchronization Hands' Movements

Aging effects on motor control function

Kyota Aoki and Hisanori Hotta
Graduate school of Engineering, Utsunomiya University, Utsunomiya, Japan
e-mail: kyota@is.utsunomiya-u.ac.jp hisanori6432@yahoo.co.jp

***Abstract*—In advanced countries, populations are getting older. Cognitive disorders are an important problem in these countries. We need to measure the deterioration of brain functions with the process of aging. For synchronizing with other motion, we need to feel the other motion, to recognize the motion, to memorize the motion, and to generate the synchronizing motion. We need many kinds of brain functions to perform a cooperative movement. The authors proposed the cooperative visual synchronization task, its' measuring method, implementation and experiments to measure and evaluate the performance of motor control functions. The new task and the measuring method enable to measure the precise movements easily and in a short period of time. The proposed method is safe, because there is no need to attach a device to a subject nor to make exaggerated motions. This paper presents a method to evaluate the aging process of motor control functions using the objective measurement of cooperative movement in both hands, its implementation, and experiments.**

Keywords-aging; aging process; motor control function; measurement; evaluation.

I. INTRODUCTION

With aging, our physical function deteriorates, and also our brain functions do. In advanced countries, our populations are getting older. Our physical deteriorations are measured easily. We need to measure the deterioration of brain functions also. There are tests to measure memory functions and congnitive disorders.

A cooperative movement with other movements is more difficult than simple movements. For instance, clapping hands is easy. However, clapping hands synchronizing with other people is difficult. Synchronizing movement is the base of cooperative movement. For synchronizing with another motion, we need to feel the other motion, to recognize the motion, to memorize the motion, and to generate the synchronizing motion. We may estimate the performance of total brain function by observing the process of synchronizing movement.

There are many motor tasks that measure human motor function abilities. They are the Purdue pegboard task, a seal affixation task, a tray carrying task, etc. [1] - [3]. These tasks estimate human motor function abilities based on the results from the tasks. There is no observation on the process of the tasks. There are also some synchronization tasks used to measure the motor function of a human. One example is a synchronization of finger taps with periodically flashing visual stimuli and synchronization with an auditory metronome. In these tasks, the timing between the stimuli and the tapping is measured. There is no observation about the process of the tapping [4] - [10].

Recently, many cheap and easy measurement methods for the movements of a human body have been developed. For instance, some of these sensors include Kinect sensor, and Leap motion sensor [11] [12]. There are many applications that use these sensors to control computers. For example, there are many video games that use these sensors to control avatars in the games [13].

Using the new motion sensor, we can measure the motion of hands easily and precisely. The human hands are the parts of the body that can make the most complex movements. We have proposed a method that measures the precise movements of hands synchronizing the movements of hands on a display. The synchronization needs visual perception of the displayed hands' images and precise control over the arm muscles. The resulting measure is very sensitive. With this measure, we can observe the performance of the motor function precisely [14].

This paper proposes a new estimation method to evaluate the performance of a brain function of an elderly person with the measurement of a motion control function in cooperative synchronizing movements. We believe this method helps to detect the cognition disorder in early stages.

The rest of this paper is organized as follows. Section II discusses the task to synchronize hands' movements with visual presentation. Next, we discuss the experimental setup in Section III, and show our experimental results in Section IV. Finally, we conclude this work in Section V.

II. VISUAL SYNCHRONIZATION TASK

There are many motor tasks that intend to measure the motor function of a human. However, most of these tasks measure the results from the tasks. There are some tasks that measure the synchronization between a finger tap and stimuli. With human observations, it is difficult to measure the process of synchronizing movements. Now, we can use a Kinect sensor and a Leap Motion sensor. These sensors measure the three-dimensional movements of a human body. With these sensors, we can measure the precise movements of a human body.

We can synchronize our movements with each other. For instance, when dancing, dancers can synchronize their movements with each other. A synchronization of movement is more difficult work than a simple imitation of movement. To generate synchronized movements, we need to observe

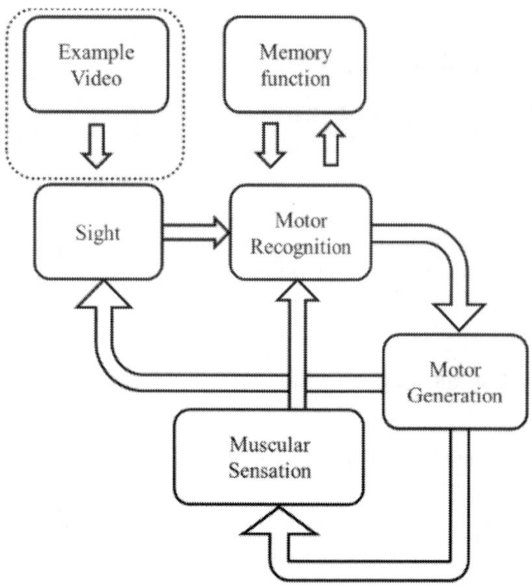

Figure 1. Relations among functions.

the motion to be synchronized. We need to generate the motion to be similar to the motion synchronized. We need to observe the generated motion synchronizing the original motion. We need to estimate the divergence between the original motion synchronized and the motion synchronizing the original motion. We need to control the speed of the motion synchronizing. These functions form a feedback loop. However, there is a delay in our processing. To compensate our brain's processing delay, we need to estimate the delay itself and make proper amount of feedforward.

This processing loop is shown in Fig. 1. For estimating the total brain function, we need to include all the functions of the brain. The visual synchronization task includes vision and motor functions. The vision includes not only the static sight, but also the dynamic sight.

The visual synchronization is more difficult than audio synchronization. So, we observe the wider brain functions with the visual synchronization tasks than the audio synchronization tasks.

Our proposed visual synchronization task is the synchronization between the position of stimuli on a display and the position of the hands. Our synchronization task is not the synchronization between the timing of the stimuli and the timing of action. The measurement of timing is only one scalar value in a cycle of stimuli. In our proposed synchronization task, the measuring result is a sequence of triples of the positions of the stimuli and the ones of subject's hands in a cycle of stimuli. For instance, we have 100 measurements in a cycle of stimuli.

A. Motion synchronization measure

We define the synchronization measure using Fast Fourier Transform (FFT) results of the estimated poses of both hands in each cycle. If a subject makes complete synchronization to the stimuli, the resulting pose of both hands follows a complete sine curve. As a result, at every cycle of the rotation of hands, the result of FFT has a zero value at the second term or higher terms. We define the measure as (1). This measure increases with the distance from ideal sine curve.

$$NSM = \left(\sum_{x=2}^{t/4} m_x\right) \Big/ m_1 \tag{1}$$

In (1), t is the number of terms. m_x is the absolute value of the x-th term of the result of FFT. m_1 is the power of the lowest frequency. This represents one cycle of a hand's rotation. If the rotation of a hand follows the stimuli images precisely, m_1 carries all powers of the hand's rotation. Other terms carry no power. In that case, the measure in (1) is 0.

m_0 is a value that represents the average of poses. This is not included in (1). As a result, this measure does not depend on the absolute poses of hands.

We call this measure as Non-Smoothness-Measure (NSM). This measure may span from 0 to infinity.

Our proposed system observes two hands. So at every cycle, we have two NSMs.

B. Phase

The NSM is the measure of the difference of a motion based on the displayed motion. However, there is a difference of timing between the displayed motion and a user's motion. The tapping test measures the difference between a stimulus and the response of a user. In the proposed synchronization task, the difference in timing is the difference of phases.

In the result of FFT, there are phases of all frequencies. In our experiments, these are from 0 Hz to 50 Hz. The signal of 1 Hz represents the ideal motion based on the proposed example motion. Therefore, we use the phase of the signal of 1 Hz for evaluating the timing of the motion.

III. Experiments and Discussions

A. Experiments Setup

From the pre-experiments, the speed of the hands' rotation is best at one cycle per second. Subjects need about three cycles to synchronize their movements of hands to the proposed motion images and remember the motion. As a result, one trial of an experiment needs at least 11 S. For getting reliable results, we decided that the length of a trial would be 25 cycles of rotations. This means that one trial needs 25 S. Fig. 2 shows the relations among parts, cycles, and sections in a trial. A cycle is one flip of hands. There are two parts. One part is an example displaying part. The other is no-example displaying part. The sections are periods to analyze measured data. The first section shows the status of a subject in the motion example displaying part. The second section does the status just after the disappearance of the motion example. The third section shows the status a few seconds after the example motion disappears. Before starting a trial, we instruct subjects to synchronize their hands to the displayed hands' images and continue to move the hands after the example motion disappears.

B. Experiment

1) Young and Healthy People

We obtained 156 valid trials with four healthy, male students, with ages between 23 - 24 years old. At each trial, we have 25 pairs of NSMs and 25 pairs of phases, at most. In many cases, a subject could not move his hands as the displayed hands at the first cycle. The NSM shows the difference of the motion of subject's hands from the proposed example motion. The phase represents the difference of the timing between the proposed example motion and the motion made by a subject.

2) Elderly people

We performed experiments with elderly people, 75 years old in average. They are all healthy for their age. In our observation, one female has difficulty walking. So, we have 14 healthy elderly people, four males and ten females. Each one made two trials, for a total of 28 trials. One trial had a failure in measurement. We obtained 27 valid measurements of the trials.

3) Measure for a trial

In a single cycle, the measured movements of hands may match the proposed example movements accidentally. We estimate the performance of the motion control function with the average motions in three continuous cycles. And, we estimate the performance of a subject in a trial with the best movements in the averages of three continuous cycles.

Equation (2) defines the performance of a hand in a trial.

$$\mathrm{NSMH} = \min_{i=1,8} average(NSM_i, NSM_{i+1}, NSM_{i+2}) \quad (2)$$

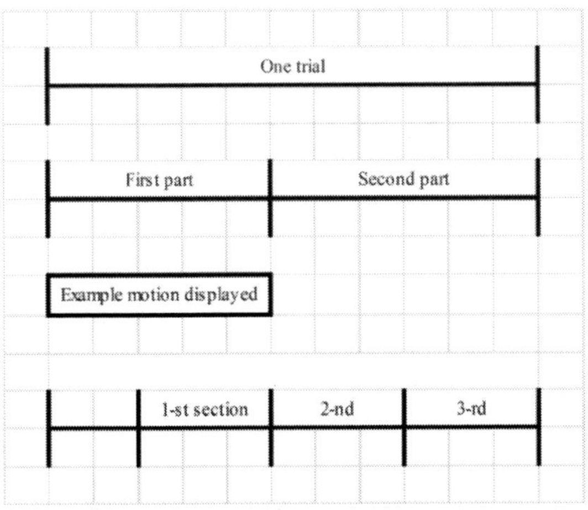

Figure 2. Relations among a session, phases and sections

NSMH is the performance of a hand in a trial. NSM_i is the NSM at i-th cycle defined as (1). We have two NSMHs in a trial. They represent the performances of both hands.

We define the performance measure in a trial as (3).

$$\mathrm{NSMT} = \min(NSMH_L, NSMH_R) \quad (3)$$

In (3), NSMT is the performance measure in a trial. $NSMH_L$ is the NSMH of the left hand. $NSMH_R$ is the NSMH of the right hand. This NSMT represents the performance of a subject in a trial.

IV. Result and Discussions

A. NSMs

1) Young people

Table I summarizes the NSMs at each cycle in young people. At the first cycle, a subject tries to synchronize his hands' motions with the displayed example motion. The average NSM of the first cycle is larger than other cycles. After three cycles, a subject completes the synchronization of his hands to the displayed motions. The NSMs at cycle 3 to cycle 10 are low. At the start of cycle 11, the example hands image disappears. The NSM at cycle 11 increases a little. The differences among cycles are small. Fig.3 shows the average of NSMs in each cycle.

In our experiments, the memory related to simple motion is good in the first five seconds from the disappearance of the

TABLE I. NSMS.OF YOUNG PEOPLE

Cycle	Example motion	Average	Standard derivation
1	Y	0.390	0.153
2	Y	0.266	0.058
3	Y	0.267	0.060
4	Y	0.253	0.054
5	Y	0.256	0.062
6	Y	0.255	0.061
7	Y	0.257	0.103
8	Y	0.252	0.056
9	Y	0.252	0.062
10	Y	0.248	0.059
11	N	0.265	0.070
12	N	0.261	0.074
13	N	0.265	0.078
14	N	0.267	0.068
15	N	0.269	0.076
16	N	0.270	0.083
17	N	0.290	0.135
18	N	0.306	0.185
19	N	0.299	0.151
20	N	0.284	0.147
21	N	0.351	0.771
22	N	0.316	0.355
23	N	0.378	1.086
24	N	0.405	1.468
25	N	0.315	0.156

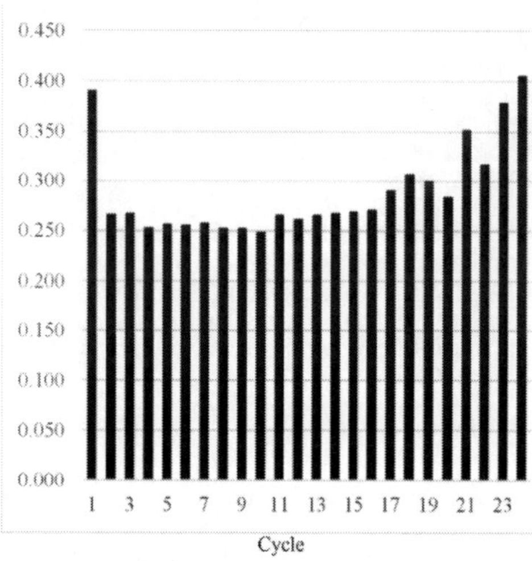

Figure 3. NSMs of each cycle in young people.

proposed example motion shown in Fig. 3. After five seconds, there is a little loss in motion precision.

We computed the difference between the distribution of the NSMs at cycle 10 and the distribution of other cycles after cycle 10.

We confirm that they have the same distributions using t-test. Table II shows the probability of sameness of the distributions from one of the cycle 10. Fig. 4 shows the probabilities. From cycle 13 to cycle 19, the probabilities are decreasing. This shows that the short-term memory of motor function decrease rapidly. After cycle 20, subjects lost the memory about the motion, and their hands' motions became more random.

With NSMTs, we estimate the performance of the younger people in synchronizing their hands' movement to the displayed hands' movement. Fig. 5 shows the distribution of NSMTs of younger subjects. The NSMTs concentrate around 0.2.

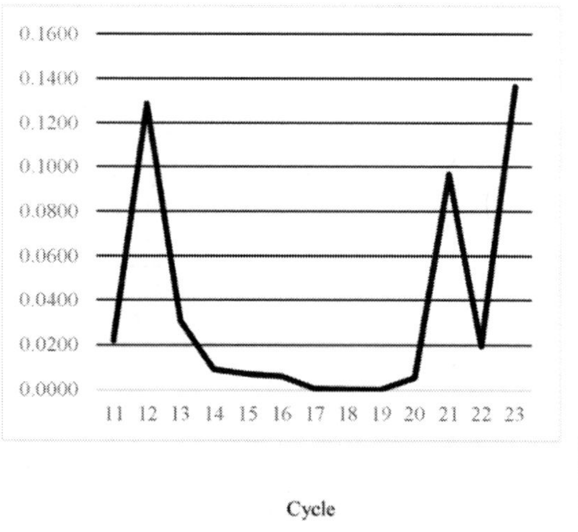

Figure 4. Probabilities of the sameness of the NSMs distributions.

2) Elderly people

Table III summarizes the NSMs at each cycle in elderly people. At the first cycle, an elderly subject synchronizes his hands' motions to the displayed example motion. The average NSM of the first cycle is larger than other cycles. After six cycles, a subject finishes to synchronize his hands to the displayed motions. The NSMs at cycle 4 to cycle 10 are low. At the start of cycle 11, the displayed example hands

TABLE II. PROBABILITIES OF SAMENESS OF NSM TO THE CYCLE 10.

Cycle	Probability of sameness
11	0.0219
12	0.1287
13	0.0305
14	0.0091
15	0.0070
16	0.0060
17	0.0005
18	0.0002
19	0.0001
20	0.0053
21	0.0966
22	0.0194
23	0.1365

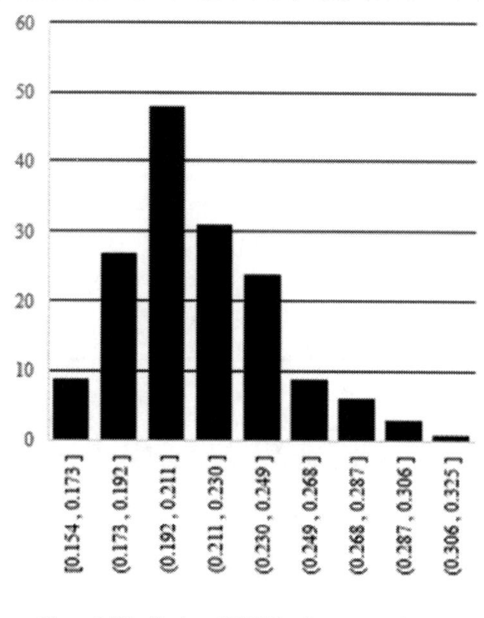

Figure 5. Distribution of NSMTs of young people.

image disappears. The NSM increases from cycle 11 to cycle 14. The differences between cycles are not large. Fig. 6 shows the average of NSMs in each cycle.

3) Comparison between young people and elderly people

We have 158 NSMTs of young people and 27 NSMTs of elderly people. On average, the NSMTs of young people are smaller than the NSMTs of elderly people. However, we need to check the reliability. We perform a t-test with these two groups of NSMTs. Table IV shows the result of the t-test. The probability of being the same is less than 10^{-9}. The deference between young people and elderly people is significant. This implies that NSMT can measure the deterioration as a result of the aging process. There is an apparent difference between NSMs of young people and ones of elderly people, as shown in Fig. 3 and Fig. 6.

TABLE III. NSMS OF ELDERLY PEOPLE

Cycle	Example motion	Average	Standard derivation
1	Y	1.474	0.953
2	Y	1.057	1.212
3	Y	0.887	1.069
4	Y	0.800	1.044
5	Y	0.645	1.016
6	Y	0.490	0.393
7	Y	0.511	0.376
8	Y	0.575	0.654
9	Y	0.438	0.218
10	Y	0.433	0.221
11	N	0.723	0.637
12	N	0.921	1.263
13	N	0.600	0.580
14	N	0.787	0.997
15	N	0.711	0.667
16	N	0.549	0.471
17	N	0.508	0.396
18	N	0.588	0.500
19	N	0.700	0.567
20	N	0.648	0.551
21	N	0.753	0.703
22	N	0.773	0.737
23	N	0.644	0.509
24	N	0.559	0.512
25	N	0.565	0.125

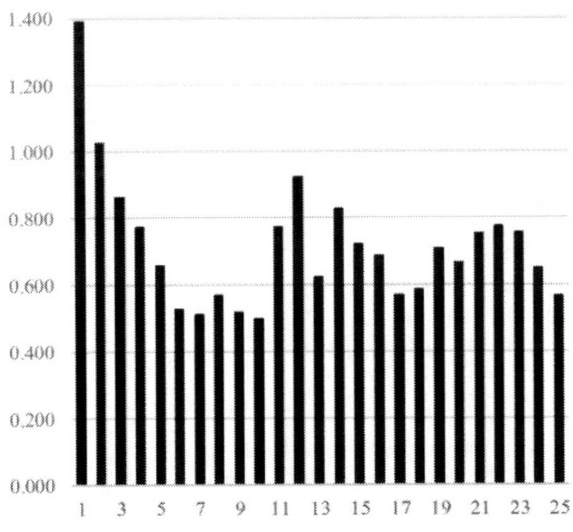

Figure 6. NSMs in each cycle on elderly people.

TABLE IV. T-TEST BETWEEN NSMTS OF YOUNG AND ELDER/

	Elder	Young
Average	0.395186	0.214459
Distribution	0.009645	0.000892
#samples	27	158
Freedom	27	
t	9.487605	
P(T<=t)	2.17E-10	
t	1.703288	
P(T<=t) Both	4.33E-10	
t both	2.051831	

4) Aging in NSMTs

In elderly people, the deterioration of motor control function increases with the aging process. Fig. 7 shows the relation between the age and the NSMT of each elderly person. The correlation coefficient between the age and the NSMT of elderly people is 0.467. There is a linear relation between the age and the NSMT. The linear approximation is (4).

$$\mathrm{NSMT} = 0.0088a - 0.26 \qquad (4)$$

In (4), NSMT is the performance measure of motor control function. a is the age of a subject. The age ranges between 66 years old and 83 years old.

The average of NSMTs is about 0.2 in young people. If the deterioration of motor control function shows a linear

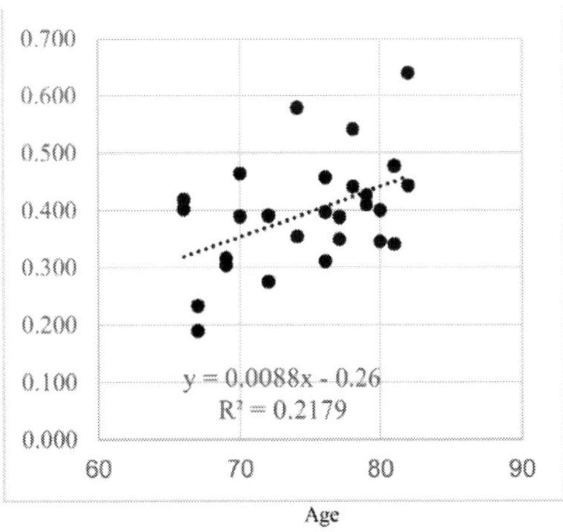

Figure 7. Age-NSMTs relation of elderly people.

relation with the age of a subject from the start of the deterioration, we can estimate the motor control function age with (5) over 53 years old people.

$$\text{NSMA} = 114\text{MSNT} + 29.5 \qquad (5)$$

In (5), NSMA is an aging years of motor control function. NSMT is a measured NSM at a trial. This shows the measurement of motor control function can estimate the aging of a brain function of elderly people.

B. *Phases*

The phase of the measured motion represents the timing of motion. In phases, there is apparent difference between the first period where the example motion is displayed and the second period where the example is not displayed in a trial. In the part from cycle 1 to cycle 10, the phases keep a similar value. From cycle 11, the phases change gradually. This represents the difference between the speed of the example motion and one of the memorized motions. From this phase change, we can measure the difference of timings between the example motion and the memorized motion.

1) Young people

We assume that the phase change in the first part of a trial is smaller than the phase change in the second part of the trial. We divided all cycles into three sections. To confirm this assumption, we calculate the linear approximation of the phases in each section. The first section starts from cycle 4, and ends at cycle10. The second section goes from cycle 11 to cycle 17. The third section goes from cycle 17 to cycle 23, as shown in Fig. 2. In 158 valid trials, there are delays and advances in phases. We evaluate phase change in absolute value.

TABLE V. PHASE CHANGES IN SECTIONS OF YOUNG PEOPLE.

Section	Cycles	Slant of phase change
1	4-10	0.022
2	11-17	0.111
3	17-23	0.113

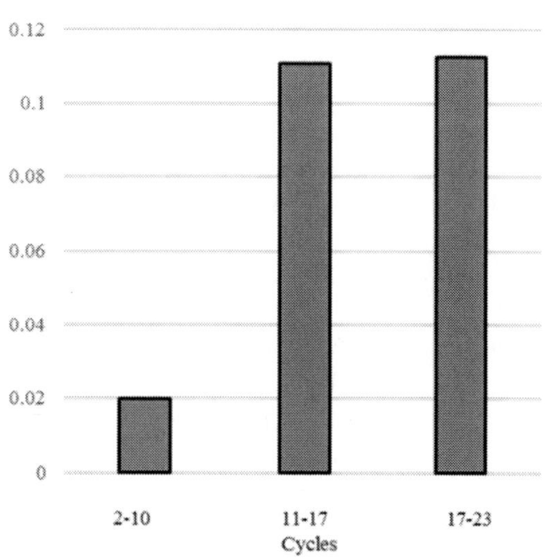

Figure 8. Phase changes in cycles.

Table V shows the averages of the slant of each section. The average absolute slant of phases in the first section is smaller than the one in the second section and the third section. Fig. 8 shows this relation. In Fig. 8, there is an apparent increase of phase changes in 2nd and 3rd sections.

Statistically, the first section and the second section have difference bases. Calculation of the t-test confirms that the difference is significant. The t-measure between these two sections is over 12. The probability is under 10^{-26}. The t-measure between the second section and the third section is 0.18. The probability is over 0.85. This confirms that the second and the third sections have a same base. This result means that the memory about the timing of motion remains for at least 15 seconds.

2) Elderly people

We also calculate the linear approximation of the phases. Table VI shows the average absolute slant of phase's change in each section.

There are apparent differences of the phase changes between young people and elderly people. Elderly people have some difficulties to keep the pace of flipping their hands. In phases, it is difficult to find the proper scale representing an aging process.

C. *Discussion*

With the NSMs, there is no apparent change between with and without a displayed motion example. Before 15 seconds, there is little decay of the memory of motion. After 16 seconds, Fig. 3 shows a little increase of the NSMs.

With the phases, there is apparent difference between with and without a displayed motion example. The changes of measured phases represent an error in the timing of a measured motion. Some trials show delay, and others show advance. The phase change shows the error about the memory of the timing. The proposed method measures the timing and the process of movements. A classical tapping test measures the timing only. However, in this experiment, the difference of 0.001 radian in phase is the difference of 0.00016 seconds in time. The proposed timing measure about motor function based on the phase of the basic movement is very keen. The classical tapping test can measure the difference of 0.0001 seconds now. However, the mechanical features about a hand and a switch make it difficult to measure the small difference of time.

V. CONCLUSION

This paper proposes the pair of the measurement and evaluation method of motor control function to estimate the deterioration with an aging process. The proposed method is implemented and tested in experiments. The task is easy to perform. For instance, it needs only 25 seconds. The

TABLE VI. PHASE CHANGES IN SECTIONS OF ELDERLY PEOPLE.

Section	Cycles	Slant of phase change
1	4-10	0.146
2	11-17	0.323
3	17-23	0.373

proposed Non-Smoothness Measure has enough power of discrimination of a motor control function. The phase changes also have enough power to measure the very small error in timing remembered.

The experimental results confirm that the proposed method can measure and evaluate the deterioration of a motor control function with an aging process precisely. This paper proposed a method to estimate the age according to the aging process of the motor control function. This age helps to measure the deterioration of the brain function, and it can detect the very first stage of cognitive impairment. We will perform larger scale experiments in the next step.

ACKNOWLEDGMENT

This work is supported with JSPS16K01057.

REFERENCES

[1] Lafayette Instrument, "Purdue Pegboard Test", http://www.lafayetteevaluation.com/product_detail.asp?ItemID=159, retrieved 2016/08/29.

[2] S. Hirata, Y. Kitajima, T. Hosobuchi ,and M. Kokubun, "The speed and accuracy of fine motor actions in children with intellectual disabilities," Bulletin of Tokyo Gakugei University, 59, pp. 263-267, 2008.

[3] M. Kokubun, "Are children with down syndrome less careful in performing a tray-carrying task than children with other types of mental retardation?" Perceptual and Motor Skills, 88:3c, pp. 1173-1176, 1999.

[4] M. J. Hove, J. R. Iversen, A. Zhang, and B. H. Repp, "Synchronization with competing visual and auditory rhythms:bouncing ball meets metronome," Psychological Research, 77, pp. 388–398, 2013.

[5] Y. Sugano, M. Keetels, and J. Vroomen, "The Build-Up and Transfer of Sensorimotor Temporal Recalibration Measured via a Synchronization Task," Front Psychol. 2012; 3: 246, doi:10.3389/fpsyg.2012.00246, 2012.

[6] V. Krause, B. Pollok, and A. Schnitzler, "Perception in action: The impact of sensory information on sensorimotor synchronization in musicians and non-musicians," Acta Psychologica, Volume 133, Issue 1, pp. 28–37, January 2010.

[7] J. Lee, J. Chai, and P. S. A. Reitsma, "Interactive Control of Avatars Animated with Human Motion Data," ACM Transactions on Graphics (TOG). ACM pp. 491-500, 2002.

[8] S. Vercruysse, J. Spildooren, E. Heremans, J. Vandenbossche, O. Levin, N. Wenderoth, S. P. Swinnen, L. Janssens, W. Vandenberghe, and A. Nieuwboer, "Freezing in Parkinson's disease: a spatiotemporal motor disorder beyond gait," Movement Disorders, 27(2), pp. 254-263, 2012

[9] M. A. Smith, J. Brandt, and R. Shadmehr, "Motor disorder in Huntington's disease begins as a dysfunction in error feedback control," Nature, 403, pp. 544-549, (3 February 2000).

[10] K. Rubia, A. B. Smith, M. J. Brammer, B. Toone, and E. Taylor, "Abnormal Brain Activation During Inhibition and Error Detection in Medication-Naïve Adolescents with ADHD," Am Psychiatric Assoc., 162, 6, pp. 1067-1075, June 2005.

[11] Microsoft, "Kinect for Windows," https://www.microsoft.com/en-us/kinectforwindows/, retrieved 2015/07/07.

[12] Leap Motion, "Leap Motion SDK," https://developer.leapmotion.com/, retrieved at 20150317.

[13] Microsoftstore, "Kinect," http://www.microsoftstore.com/store/msusa/en_US/list/Kinect/categoryID.64752700, retrieved 2015/07/08.

[14] K. Aoki, and H. Hotta, "Brain Activity Estimation with Precise Motor Measurements of Visual Synchronization Task of Hands", 2015, Global Health 2015, pp.39-44, ISBN: 978-1-61208-424-4, ISSN: 2308-4553, Nice, France, July 2015.

Wearable Inertial Sensors as a Tool for Quantitative Assessment of Progress during Rehabilitation

Salvatore Tedesco[1], Andrea Urru[1], James Peckitt[2], Brendan O'Flynn[1]
[1]WSN Group, Micro & Nano Systems - Tyndall National Institute
[2] Mardyke Arena Clinic
University College Cork
Cork, Ireland
e-mail: salvatore.tedesco@tyndall.ie

Abstract— **Biomechanics analysis is frequently used in both clinical and sporting practice in order to assess human motion and performance of defined tasks. Whilst camera-based motion systems have long been regarded as the 'Gold-standard' for quantitative movement-based analysis, their application is not without limitations as regards potential sources of variability in measurements, high costs, and practicality of use for larger patient/subject groups. Another more practical approach, which presents itself as a viable solution to biomechanical motion capture and monitoring in sporting and patient groups, is through the use of small-size low-cost wearable Micro-ElectroMechanical Systems (MEMs)-based inertial sensors. The clinical aim of the present work is to evaluate gait during rehabilitation following knee injuries and to identify gait abnormalities through a wireless inertial sensing system. This system was developed at the Tyndall National Institute to meet clinician-defined needs, and is able to provide a complete biomechanics assessment without the constraints of a motion capture laboratory. The derived motion parameter outcomes can be analyzed by clinicians and sport scientists to study the overall patients' condition and provide accurate medical feedback as to their rehabilitative progress. Detection of atypical movement characteristics is possible by comparing the performance and variability in motion characteristics in the patient's affected and unaffected lower-limbs. The work is ongoing, and to date the system has been tested on only one impaired subject, additional clinical trials are currently being planned with an enhanced number of injured subjects. This will provide a more robust statistical analysis of the data in the study. The present feasibility study proved that inertial sensors can be used for a quantitative assessment of knee joint mobility, and gait mechanics during the rehabilitation program of injured subjects and can provide valuable information to clinical experts as regards patient rehabilitation.**

Keywords- Inertial Sensors; Wearable Microsystems; Rehabilitation.

I. Introduction

Biomechanics analysis is frequently used in both clinical and sporting practice by clinicians to track patient progress and to define rehabilitations programs through the assessment of human motion during the performance of defined tasks. Common examples include the use of camera-based motion analysis systems during formal gait analysis by rehabilitation professionals to ascertain measurements of Temporal (Time) & Spatial Characteristics associated with gait parameters. This enables clinicians to identify gait deviations in paediatric and amputee populations, screening elderly people for risk of falling, to objectively monitor patient's progress, and to help determine the efficacy of surgical and therapy interventions [1]-[4].

Whilst camera-based motion systems have long been regarded as the 'Gold-standard' of quantitative movement-based analysis, their application is not without limitations as regards potential sources of variability in measurements, relatively high costs of instrumentation including access to specialist motion labs, as well as practically of application for larger patient/subject groups, as has been discussed by Chau et al. [5].

From a clinical perspective, observational forms of clinical gait analysis frequently forms the corner stone of patient knee joint assessment, and is typically used in parallel with manual clinical assessment techniques, such as stress-testing evaluation of joint laxity, range of movement (ROM), and manual and/or isokinetic strength assessment, as well as contextual subjective patient questionnaires, such as International Knee Documentation Committee (Form) IKDC [1][3].

However, the use of observational gait analysis (non-empirical assessments), even when used by experienced clinicians, may not be adequate or sensitive enough to detect subtle clinical pathological changes in movement following knee surgery [3][4].

Another approach, which has been explored as a more practical and viable solution to biomechanical motion capture and monitoring in sporting and patient groups, is through the use of small-size low-cost wearable inertial sensors [1][3][4].

Figure 1. Tyndall Wireless Inertial Measurement Unit (WIMU)

Nevertheless, despite the great amount of work presented in literature on inertial sensors for biomechanics, such a technology has been adopted for monitoring lower-limbs during rehabilitation or tele-rehabilitation only in few cases (for example [6][7]). In even fewer cases, inertial sensors were adopted to assess injured athletes' joint movement during rehab (such as, ankle [8] or shoulder [9]). To the best of authors' knowledge, only one study [10] has investigated athletes' movement following knee injuries; however, the proposed solution (e.g., a full-body suit equipped with 10 wearable sensors) may be cumbersome.

The clinical aim of the present ongoing work is to evaluate gait during rehabilitation following knee injuries and to identify gait abnormalities through a wireless portable easy-to-use inertial sensing system. The wireless sensing system, developed at the Tyndall National Institute, consists of two sensors per limb, and is able to provide a complete biomechanics assessment (without the constraints of a laboratory) for a series of scripted activities. The derived outcome can be analyzed by clinicians and sport scientists to study the overall patients' condition and provide accurate medical feedback, thus proving that inertial sensors can be used for a quantitative assessment of knee joint mobility, and gait mechanics during the rehabilitation program of injured athletes.

This work was a phase one feasibility study. In order to further validate the drawn conclusions in statistical terms, additional clinical trials, with larger and homogeneous populations, are needed and currently being planned. This feasibility study represent a prerequisite to a larger cross-sectional/cohort trial that will assess sport performance analysis and movement pattern alterations detection in people following knee injuries pathologies through wearable inertial sensing technology. Ethics approval has been secured for this proof-of-concept work and larger scale validation trials.

The present work is organized as follows. The methodology of the study, with a description of the hardware and of the protocol used during the test, are described in Section II. The obtained results are shown in Section III and exhaustively analyzed and discussed, also providing new requirements for future clinical trials. Finally, conclusions are drawn in the last section.

II. Methodology, Hardware, and Protocol

The parameters taken into account for a complete assessment are as follows:

- Temporal events: toe-offs, heel-strikes, mid-stance;
- Temporal intervals: gait cycle duration, stance phase, swing phase, single and double support, cadence (or step rate), number of cycles, swing symmetry;
- Spatial parameters: stride length, stride velocity (or speed), peak angular velocity, shank clearance;
- Knee range of motion.

For each of those parameters, it is possible to calculate min, max, mean, median, standard deviation values and extrapolate the related Coefficient of Variability (CV).

The system consists of two Tyndall Wireless Inertial Measurement Units (WIMUs) per leg with 3D accelerometer/gyro (@ 400 Hz) and Bluetooth Low-Energy/SD cards (Fig. 1). WIMUs have been attached to the anterior tibia, 10 cm below the tibial tuberosity, and to the lateral thigh, 15 cm above the tibial tuberosity using surgical adhesive tape.

The data fusion algorithms are implemented in Matlab, and the scenarios considered are walking and hamstring curl – defined by physiotherapists as good indicators of rehabilitation progress. In the walking scenario, the subject stands on a treadmill, which is then being operated at different speeds (3 and 4 km/h) for approximately one minute per test. In the hamstring curl scenario, the subject stands and bends the knee raising the heel toward the ceiling as far as possible without pain, relaxing the leg after each repetition. A significant number of repetitions for each scenario was carried out, so as to provide an accurate picture of the conditions. The system has been tested with an impaired subject. The impaired subject is a female athlete, age: 44, height: 161 cm, and weight: 52 kg, with good general health status, with history of knee injuries in the last 2 months before testing (reconstructed anterior cruciate ligament in the left leg following a sporting injury), and tested for 3 months starting from one month after surgery.

III. Results and Discussion

Knee joint angles have been estimated for the participant. Results are shown for the hamstring curl scenario in Fig. 2 (left leg on the left, and right leg on the right), where each line indicates the reference joint angle values extrapolated from the exercises performed by the unimpaired subject at 4, 6, 8, and 10 weeks following the knee surgery. Those lines represent the average characteristic of all the individual repetitions carried out at each testing session. While the lines for the right lower-limb are always consistent throughout the rehabilitation process, the left impaired lower-limb shows a great difference in the results obtained in the first session (after 4 weeks) compared to the other ones. All the following sessions are comparable also for the left lower-limb.

Finally, all the gait spatio-temporal parameters mentioned in Section 2 have been calculated for the walking scenario (at 3 and 4 km/h) for both legs of the impaired subject. Results are summarized in Fig. 3.

As per the temporal variables, the first session after 4 weeks shows a strong difference between left and right leg due to injury's effects on gait, especially highlighted by the CV. In the following testing sessions, performance from affected and unaffected lower-limbs become comparable. Those considerations are valid for both speeds of the walking scenario.

Dissimilarities are also clear as per spatial parameters.

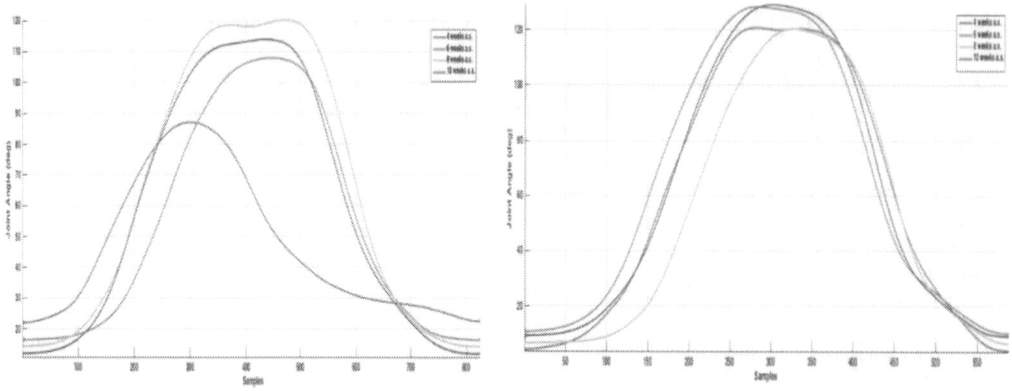

Figure 2. Joint Angles for left leg (left) and right leg (right) measured throughout rehabilitation (hamstring curl exercise).

	4 weeks a.s.		6 weeks a.s.		8 weeks a.s.		10 weeks a.s.	
	Left Leg	Right Leg	Left Leg	Right Leg	Left Leg	Right Leg	Left Leg	Right Leg
Gait Cycle Duration (s)	1.3264 ± 0.0540 (4.0707)	1.3275 ± 0.0188 (1.4184)	1.4049 ± 0.0166 (1.1806)	1.4030 ± 0.0179 (1.2766)	1.4053 ± 0.0196 (1.3976)	1.4092 ± 0.0198 (1.4031)	1.3572 ± 0.0224 (1.6472)	1.3587 ± 0.0194 (1.4310)
Stance Phase (s)	0.8535 ± 0.0471 (5.5159)	0.8560 ± 0.0144 (1.6859)	0.8828 ± 0.0130 (1.4753)	0.9121 ± 0.0153 (1.6726)	0.8942 ± 0.0204 (2.2778)	0.9171 ± 0.0138 (1.5095)	0.8541 ± 0.0224 (2.6248)	0.8957 ± 0.0173 (1.9320)
Swing Phase (s)	0.4730 ± 0.0489 (10.3425)	0.4715 ± 0.0101 (2.1442)	0.5221 ± 0.0124 (2.3671)	0.4908 ± 0.0109 (2.2123)	0.5111 ± 0.0071 (1.3962)	0.4922 ± 0.0128 (2.6004)	0.5031 ± 0.0152 (3.0247)	0.4630 ± 0.0087 (1.8799)
Single Support (s)	0.9444 ± 0.0523 (5.5378)		1.0129 ± 0.0168 (1.6558)		1.0032 ± 0.0130 (1.2921)		0.9662 ± 0.0191 (1.9744)	
Double Support (s)	0.3830 ± 0.0511 (13.3479)		0.3920 ± 0.0144 (3.6648)		0.4060 ± 0.0152 (3.7526)		0.3926 ± 0.0230 (5.8471)	
Cadence (step/min)	45.2346	45.1977	42.7082	42.7656	42.6959	42.5762	44.2072	44.1590
Swing Simmetry	1.0076 ± 0.1056 (10.4839)		0.9406 ± 0.0293 (3.1191)		0.9633 ± 0.0313 (3.2443)		0.9210 ± 0.0278 (3.0197)	
Stride Length (m)	0.9218 ± 0.1436 (15.5630)	1.2633 ± 0.0717 (5.6783)	1.0038 ± 0.0693 (6.9074)	1.0881 ± 0.0744 (6.8416)	0.9514 ± 0.0614 (6.4526)	1.1749 ± 0.0960 (8.1704)	1.0061 ± 0.0932 (9.2608)	1.0271 ± 0.0670 (6.5214)
Stride Speed (m/s)	0.6953 ± 0.1067 (15.3439)	0.9517 ± 0.0541 (5.6878)	0.7144 ± 0.0474 (6.6413)	0.7754 ± 0.0492 (6.3493)	0.6770 ± 0.0421 (6.2134)	0.8335 ± 0.0651 (7.8095)	0.7410 ± 0.0643 (8.6810)	0.7560 ± 0.0488 (6.4557)
Shank Clearance (m)	0.0242 ± 0.0049 (20.2683)	0.0645 ± 0.0126 (19.6077)	0.0217 ± 0.0050 (23.1823)	0.0346 ± 0.0070 (20.3285)	0.0175 ± 0.0035 (19.8007)	0.0285 ± 0.0081 (28.5536)	0.0404 ± 0.0077 (19.1704)	0.0343 ± 0.0065 (19.0170)

	4 weeks a.s.		6 weeks a.s.		8 weeks a.s.		10 weeks a.s.	
	Left Leg	Right Leg	Left Leg	Right Leg	Left Leg	Right Leg	Left Leg	Right Leg
Gait Cycle Duration (s)	1.1782 ± 0.0760 (6.4469)	1.1874 ± 0.0169 (1.4195)	1.2156 ± 0.0158 (1.3031)	1.2156 ± 0.0177 (1.4587)	1.1579 ± 0.0305 (2.6329)	1.1552 ± 0.0134 (1.1590)	1.1556 ± 0.0146 (1.2609)	1.1532 ± 0.0169 (1.4726)
Stance Phase (s)	0.7734 ± 0.0532 (6.8757)	0.7369 ± 0.0140 (1.9010)	0.7647 ± 0.0155 (1.8014)	0.7438 ± 0.0134 (2.0329)	0.7070 ± 0.0206 (2.9096)	0.7217 ± 0.0116 (1.6060)	0.7092 ± 0.0128 (1.8086)	0.7254 ± 0.0125 (1.7251)
Swing Phase (s)	0.4048 ± 0.0613 (15.1344)	0.4504 ± 0.0096 (2.1377)	0.4719 ± 0.0060 (1.2652)	0.4509 ± 0.0096 (2.1247)	0.4509 ± 0.0201 (4.4652)	0.4335 ± 0.0088 (2.0206)	0.4464 ± 0.0063 (1.4001)	0.4278 ± 0.0093 (2.1727)
Single Support (s)	0.8552 ± 0.0609 (7.1259)		0.9228 ± 0.0115 (1.2477)		0.8844 ± 0.0243 (2.7436)		0.8742 ± 0.0120 (1.3770)	
Double Support (s)	0.3230 ± 0.0518 (16.0521)		0.2928 ± 0.0159 (5.4296)		0.2735 ± 0.0221 (8.0865)		0.2813 ± 0.0107 (3.8179)	
Cadence (step/min)	50.9261	50.5320	49.3577	49.3573	51.8167	51.9397	51.9220	52.0280
Swing Simmetry	1.1414 ± 0.1928 (16.8902)		0.9560 ± 0.0231 (2.4184)		0.9627 ± 0.0373 (3.8788)		0.9586 ± 0.0229 (2.3890)	
Stride Length (m)	1.0786 ± 0.0893 (8.2798)	1.3574 ± 0.0484 (3.5631)	1.0620 ± 0.0450 (4.2325)	1.2122 ± 0.0828 (6.8283)	0.9792 ± 0.1695 (17.3061)	1.2087 ± 0.0406 (3.3598)	1.1661 ± 0.0495 (4.2479)	1.1794 ± 0.0851 (7.2171)
Stride Speed (m/s)	0.9175 ± 0.0778 (8.4793)	1.1431 ± 0.0368 (3.2216)	0.8738 ± 0.0393 (4.5020)	0.9974 ± 0.0710 (7.1193)	0.8457 ± 0.1467 (17.3504)	1.0463 ± 0.0333 (3.1800)	1.0090 ± 0.0376 (3.7263)	1.0224 ± 0.0673 (6.5763)
Shank Clearance (m)	0.0277 ± 0.0051 (18.3936)	0.0553 ± 0.0080 (14.4414)	0.0255 ± 0.0033 (13.0101)	0.0502 ± 0.0132 (26.3869)	0.0281 ± 0.0053 (18.9419)	0.0499 ± 0.0144 (28.8038)	0.0411 ± 0.0052 (12.5986)	0.0469 ± 0.0079 (16.8087)

Figure 3. Walking spatio-temporal parameters at 3 (up) and 4 (down) km/h for each testing session. Mean and St. Dev. shown. CV is indicated in brackets

For instance, the evident difference (roughly 30 cm) between left and right stride length in the injured subject at 3 and 4 km/h in the first session is much larger compared to the same divergence measured during the remaining testing sessions. Indeed, the left stride length for the impaired athlete is always shorter compared to her right stride length at every speed and for every session.

The CV associated to the stride length is an additional parameter that further shows this dissimilarity. The average CV for the right stride length is always consistent at each session, whilst this is not evident in the CV measured on the subject's left lower-limb, especially on the session after 4 weeks.

The same characteristics are observed in the estimation of stride speeds and shank clearance. The difference between left and right stride speed in the following sessions is much limited compared to the same variable measured during the first one. The associated CV for right and left stride speed proves this conclusion as well.

Finally, while the difference between the mean values of left and right shank clearance is consistent at the first session at 3 km/h, and is limited in the following weeks, such a difference remain visible until the 4th session (after 10 weeks) when considering the speed at 4 km/h, which involves higher dynamic movements.

All those results indicate that the gait measurements gathered from the subject after 6 weeks are far more repeatable and stable than the gait variables collected at the first session. Therefore, this feasibility study has proved that the studied wearable inertial sensing technology is able to potentially detect atypical gait movement characteristics accurately and reliably by comparing performance and differences in the affected and unaffected lower-limbs.

Those results also proved how critical and important the first 6 weeks of rehabilitation after knee surgery may be, which should be targeted and analyzed with much frequent data captures in the following studies. Moreover, it has been shown how only joint angles or gait spatio-temporal parameters may be too limited in order to provide a complete picture of the subject's condition after the first phase of rehabilitation. Therefore, new variables should be considered so as to define also more subtle aspects of the gait change, such as postural sway, the energy expenditure and the movements' smoothness.

Finally, given that alongside re-education of motor patterning, muscular reconditioning is an important rehabilitation goal during the restoration of function after injury, further exercises targeting fatigue and muscle strength/power estimation should be included into the protocol for the following studies, so as to have a better understanding of the overall patient's progress throughout the procedure.

IV. Conclusions & Future Work

This work presented a wearable inertial system for an objective assessment of lower-limbs. Detection of atypical movement characteristics was measured by comparing performance and differences in the affected and unaffected lower-limb. The test subject will continue to be monitored throughout the complete rehabilitation in order to measure her response to therapeutic treatment. An enhanced number of subjects, with homogeneous characteristics, will also be tested in the future so as to have a more robust base for the study and further validate the drawn conclusions in statistical terms. These additional clinical trials are under development.

However, the present feasibility study proved that inertial sensors can be used for a quantitative assessment of knee joint mobility, and gait mechanics during the rehabilitation program of injured subjects.

Acknowledgments

This publication has emanated from research supported in part by a research grant from Science Foundation Ireland (SFI) and is co-funded under the European Regional Development Fund under Grant Number 13/RC/2077 – CONNECT as well as the European funded project SAFESENS under the ENIAC Program in association with Enterprise Ireland (IR20140024).

References

[1] N. Hagemeister, L. H. Yahia, N. Duval, and J. A. de Guise, ''In vivo reproducibility of a new non-invasive diagnostic tool for three-dimensional knee evaluation,'' The Knee, vol. 6, no. 3, pp. 175-181, Aug. 1999.

[2] J. W. Youdas, et al., ''Agreement between the GAITRite walkway system and a stopwatch–footfall count method for measurement of temporal and spatial gait parameters," Arch Phys Med Rehabil, vol. 87, no. 12, pp. 1648-1652, 2006.

[3] S. Lustig, R. A. Magnussen, L. Cheze, and P. Neyret, ''The KneeKG system: a review of the literature,'' Knee Surg Sports Traumatol Arthrosc., vol. 20, no. 4, pp. 633-638, 2012.

[4] B. Shabani, et al., ''Gait knee kinematics after ACL reconstruction: 3D assessment,'' International Orthopaedics (SICOT), vol. 39, no. 6, pp. 1187-1193, June 2015.

[5] T. Chau, S. Young, and S. Redekop, "Managing variability in the summary and comparison of gait data," J Neuroeng Rehabil., vol. 2, no. 22, pp. 1-20, 2005.

[6] O. M. Giggins, K. T. Sweeney, and B. Caulfield, "Rehabilitation exercise assessment using inertial sensors: a cross-sectional analytical study," J Neuroeng Rehabil., vol. 11, no. 158, pp. 1-10, 2014.

[7] R. Nerino, et al., "An improved solution for knee rehabilitation at home," 9th Int Conf Body Area Networks, pp. 62-68, 2014.

[8] S. Zhang, F. Naghdy, D. Stirling, M. Ros, and A. Gray, "Ankle injury assessment using inertial 3D data," IEEE/ASME Int Conf Adv Intell Mech, pp. 810-815, 2013.

[9] C. Tranquilli, A. Bernetti, and P. Picerno, "Ambulatory joint mobility and muscle strength assessment during rehabilitation using a single wearable inertial sensor," Medicina dello Sport, vol. 66, no. 4, pp. 583-597, Dec. 2013.

[10] D. Fitzgerald, et al., "Development of a wearable motion capture suit and virtual reality biofeedback system for the instruction and analysis of sports rehabilitation exercises," 29th Ann Int Conf IEEE EMBS, pp. 4870-4874, Aug. 2007.

A Relational Method for Determining Eventual Causality in Electronic Health Records

Bilal El-Hajj-Diab
School of Computer Science and Software Engineering
University of Wollongong
Wollongong, Australia
email: b.diab@alumni.uottawa.ca

Ali Hage-Diab
Department of Biomedical Engineering
Lebanese International University
Beirut, Lebanon
email: ali.hagediab@liu.edu.lb

Bassam Hussein
Department of Industrial Engineering
Lebanese International University
Beirut, Lebanon
email: bassam.hussein@liu.edu.lb

Mohamad Raad
Department of Computer and Communication Engineering
Lebanese International University
Beirut, Lebanon
email: mohamad.raad@liu.edu.lb

Hassan M. Khachfe
Center for Quality Assurance, Institutional Assessment & Scientific Research (QAIASR)
Lebanese International University
Beirut, Lebanon
email: hassan.khachfe@liu.edu.lb

Abstract— **The use of electronic health records (EHR) has been increased substantially to improve quality of healthcare outcomes. That's why recent EHR systems have been eagerly evolving from patient documentation systems toward intellectual tools for physicians to accomplish their tasks. Such evolution would mean continuous changes to the current EHR systems, which would be a real challenge since the designed software for these systems is usually customized based on prior requirements and thus having the correct workflow design is an essential key to allow system intellectuality and system automation. Such upgrade would give more realistic view to the patient medical profile by associating all related medical entities in a hierarchical order without losing the usability and performance of these systems. This new approach treats the EHR system as a storage manager, which is an important advantage from the development prospective especially for any future changes in any of the different layers.**

Keywords-workflow; electronic health record; causality; logical modeling.

I. INTRODUCTION

Consider a simple chat regarding a missing boy; the mother who is worried about her child not coming home sends a message to her friend asking "my boy is missing" (message 1), then the mom sends another message stating that her boy is home and no need to worry (message 2); her friend replies saying "thanks god" (message 3). So for these three messages it's obvious that "message 3" was caused by " message 2" and the correct workflow of causality of these messages concluded that the "thanks god" message makes sense since now the boy is safe and she is happy about it. Now, assume that for some reason "message 2" is missing then the whole concept would change and therefore "message 3", which is "thanks god" would look as if it was caused by the "message 1" and in this case the message would be completely misinterpreted because "thanks god" would be an answer to "my boy is missing" message.

In medical profiling, similar scenarios may apply but in much more serious conditions [1]. Consider a patient who had a minor head injury two days before going to a clinic for a severe swelling sinus and headache. After several examinations by the physician, he was diagnosed with flu. The head injury was not considered since it had nothing to do with his symptoms so it was ignored. After few years, the same patient was taken to the clinic again with serious symptoms of amnesia. After careful examination, the doctor concluded that this patient might be developing Alzheimer disease. However, there could be a possibility that the minor head injury which occurred a couple of years ago had something to do with this patient memory loss, but since this fact was not eliminated before, neither the doctor nor the patient would even consider such possibility. The flow of the symptoms and diagnosis can be thought of as messages or processes of causality. For example, head injury is "message 1", then flu symptoms is "message 2", followed by amnesia, which is "message 3". Therefore, if we had a knowledgebase to remove message 2 from the flow, then it would have been clear to the physician that amnesia (message 3) might have been caused by the head injury (message 1).

In this paper, we introduce a causal workflow protocol, which will act as a shim. A shim is a library that transparently intercepts application program interface (API) calls on top of the electronic health records (EHR) system and produces symptoms, as well as diagnoses based on causality to assist physicians to have as much as possible accurate presentation of the patient's electronic profile.

Therefore, a bolt-on causal workflow layer to be added on top of a general purpose EHR system will be responsible to enforce the constraints for association and relation between the patient medical entities and would leave all other functionalities of the EHR management intact. The bolt-on will act as a special task middle layer between the user and the system by dealing with EHR system as a storage system in most of the cases. Medical diagnosis information processing dates back to the early seventies [2][3].

Section II outlines the method and implementation procedure. Section III presents enhancement recommendations for future work. Section III concludes the article.

II. METHOD AND IMPLEMENTATION

In general, any typical medical procedure would consist of a set of sequential examinations in order to determine the diagnoses [4]. We can think of these procedures in EHR system as top to bottom hierarchical workflow tree (t) while every process or examination would be considered a node (n) in this tree of flow since every visit (v) would be considered a set of examination as illustrated in Fig. 1. Therefore, every patient visit would create a new tree of examination workflow consisting of a set of processes that may be viewed as nodes.

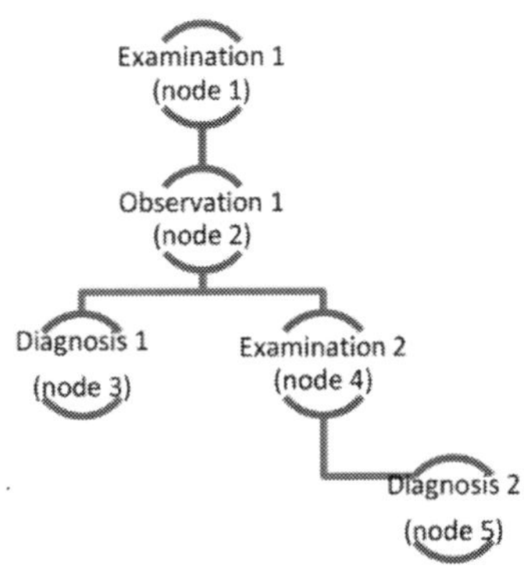

Figure 1. Patient visit process workflow.

Where the node n denotes a medical or process that can affect the decision of the practitioner and therefore would affect the process tree. Now, for m visits, a patient would have:

$\mathbf{tv_1 \ldots tv_m}$
of workflow trees
where each tree consists of
$\mathbf{n_1 \rightarrow n_2 \rightarrow n_3 \rightarrow \ldots}$

Noting the sequential relation between the nodes in every workflow, almost every patient visit to a doctor can be related to a previous condition from previous visit. Hence, there would be a cross relation between the workflow trees such that $tv_1 \rightarrow tv_2 \rightarrow \ldots$ or in another word causal relationship.

Clearly, we can see that with a missing causal reading consistency in the EHR system, the patient workflow trees would not provide the exact history and causality and therefore enormous work would be up to the physician to handle manually, redundantly, or even worse to build an assumption based on missing relations. This becomes an important operational problem [5].

The bolt-on would have customized consistency rules considered as meta-data. It would always put a unified causal tree of related processes, which the client can only see without further development. This shim would check constraint on write and read based on causal-relation of any process or examination, which in turn would guarantee having all sequential nodes returned even across different trees (horizontally in same workflow and vertically across other workflows) and provide them seamlessly and cohesively as depicted in Fig. 2. Finally, the architecture suggests layers separation of safety and live-ness concerns. So while the shim layer would handle consistency and visibility, the underlying EHR would handle the live-ness and usability to the end user.

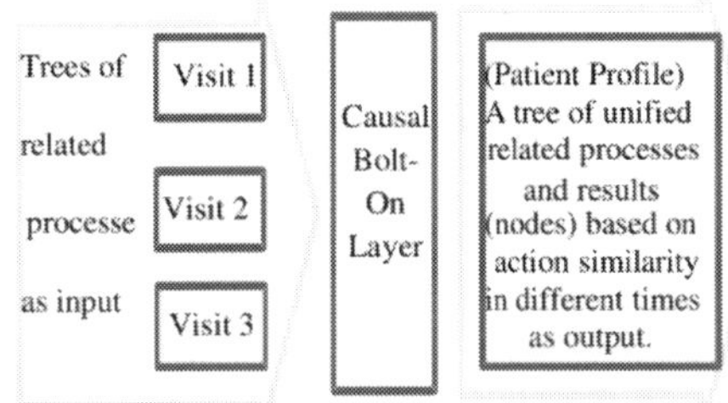

Figure 2. Bolt-on design unified causal tree.

After each patient visit, physicians would only be concerned about the diagnosis of the patient in that specific event and therefore all symptoms and incidents leading to the conclusion to a certain diagnosis would be ignored. As a result, symptoms and incidents would be observed only at the analysis level currently; however, the set of concluded diagnosis might be observed through any particular visit, although sequential observation is most common.

As shown in the above example, it's clear that valuable incidents or symptoms (process) could be overlooked in the time line of patient profile. Keep ALIVE methodology can be applied to incidents, which are candidates to be needed later. The need for each incident can be decided based on a scale from 1 to 10 concluded from a knowledge base providing the most-likability reaction between an incident, a symptom and a diagnose.

In order to better evaluate the process, we can convert the given example to a logical model. Logical models are actionable plans, strategies or maps with clear outcomes and explicit steps for solving program problems. We use the logical model in planning, implementation, evaluation and communication. A more advanced notion of such models, such as fuzzy logical models has been proposed. They have contributed to building medical expert systems to assist physicians in medical decision-making and patient care [6].

In any particular patient examination or visit, the current symptoms and incidents will be treated as inputs, observations checkups and examinations as activities where diagnosis and final conclusions are the outputs. In addition the previous diagnosis which is the history of the patient would be considered as inputs also since it is pre-given and has a direct effect on the whole logical process.

The given example can be logically proposed in the logical model in Table I.

TABLE I. PATIENT FIRST VISIT LOGICAL MODELING.

Input	Activity	Output
- Current Symptoms o Headache o Sinus - Current Incidents o Head Injury - Previous Diagnosis	- Observation - Check-up - Examination	- Diagnosis o Flu - Prescriptions o Med1 o Med 2

In the second patient visit, the logical model would most likely look as shown in Table II.

TABLE II. PATIENT SECOND VISIT LOGICAL MODELING.

Input	Activity	Output
- Current Symptoms o Headache o Amnesia - Current Incidents o <u>None</u> - Previous Diagnosis	- Observation - Check-up - Examination	- Diagnosis o Alzheimer - Prescriptions o Med 1 o Med 2 o Med 3

Consider a given knowledgebase k1: Head Injury -> Brain bleeding -> Amnesia -> …

When a patient comes back with a symptom that belongs to a knowledge base and also a child of previously inspected incident, then that incident would be flagged as ALIVE in the database and it would then be added automatically as input to the current visit logical model. Therefore, the auto-adjust logical model would be shown as illustrated in Table III.

TABLE III. PATIENT DIAGNOSIS LOGICAL MODELING WITH CAUSALITY ASSOCIATION.

Input	Activity	Output
- Current Symptoms o Headache o <u>Amnesia</u> - Current Incidents o None - **<u>ALIVE Incidents</u>** o **Minor Head Injury** - Previous Diagnosis	- Observation - Check-up - Examination	- Diagnosis o Alzheimer - Prescriptions o Med 1 o Med 2 o Med 3

The suggested association would always be present as input based on the given medical knowledge base, which in turn would provide enough resources to accurately provide the patient present condition auto-linked with all previous related conditions or incidents denoted in table II underlined and in bold in the input section.

From an implementation prospective, this process requires a set of local resources such as linked knowledge base. However with the tremendous medical data available online now a day, making certain online resources in a distributed environment available as inputs to EHR input or activities sections in the logical model would clearly enrich the system and would keep practitioners on the track with the latest medical trends. That being said, it would be beneficial to encapsulate the implemented code in the form of web services. Creating web services would easily be used by different EHR systems written indifferent programming languages and would act as a standalone layer [7]. Web services can be described as applications which are dynamically self-contained, it can easily published, consumed and located from distributed applications or models over a local network or web based or even in a local client server applications. Web services are built on top of open standards such as TCP/IP (Transmission Control Protocol/Internet Protocol), HTTP (Hypertext Transfer Protocol), also Java in-addition to HTML (HyperText Markup Language), and XML (Extensible Markup Language).

Typically, web services are XML-based information exchange systems, which use direct end-to-end application interaction by exchanging XML messages and can preserve both security and performance [8].

III. FUTURE WORK

Adding ontology to the syntactical data entery would not just ease practioner work, but also would give the possibiltiy to auto analyze medical profiles and allow sophiscticated relations between various entities to be associated [9]. Such critieria would even enhance the medical systems at global bases where signs of epidemics would be detected at earlier stages. An ontology is defined as a formal, explicit specification of a shared conceptualization [10]. Several ontological frameworks for describing space and spatial relations have been developed recently [11].

In order to develop such add-on, natural language programming (NLP) would need to be applied. NLP is a component of artificial intelligence which gives the ability of a computer program to understand human speech as it is spoken. However, this is a very challenging approach especially that human speech is not always precise. In fact, it is often ambiguous and the linguistic structure can depend on many complex variables, including slang, regional dialects and social context. Due to the challenges of NLP, we aim to create a custom local grammar that has "high" end-user with XML based easy customization at the development level, which would allow adjustments with minimal coding.

IV. CONCLUSION

Our causal bolt-on will treat EHR system as a storage manager and that would be similar to creating a 3-tears application which is an important advantage from the development prospective especially for any future changes in any of the different layers. As a result, such improvement would guarantee seamless and yet very effective addition to the electronic health records regardless of the used platform toward an optimized electronic patient profile.

REFERENCES

[1] B. El-Hajj-Diab, I. Zayour, and A. Hage-Diab, "Auto-Learning Electronic Medical Record System", 20th LAAS International Science Conference, Beirut, Lebanon, March 27–29, 2014.

[2] P. Wortman, "Medical diagnosis: an information-processing approach." Computers and Biomedical Research, pp. 315-328, 1972.

[3] E. Shortliffe and B. Buchanan, "A model of inexact reasoning in medicine." Mathematical Biosciences, pp. 351-379, 1975.

[4] S. Kumar and K. Aldrich, "Overcoming barriers to electronic medical record (EMR) implementation in the US healthcare system: A comparative study", Health Informatics Journal, pp. 306-318, December 2010.

[5] M. Öztürké, A. Tsoukiàs, and P. Vincke, "Preference modelling", In Multiple criteria decision analysis: State of the art surveys, pp. 27-59. Springer New York, 2005.

[6] R. Seising, "From vagueness in medical thought to the foundations of fuzzy reasoning in medical diagnosis", Artificial Intelligence in Medicine, pp. 237-256, 2006.

[7] I. Zayour and H. Hajjdiab, "How much integrated development environments (ides) improve productivity?", Journal of Software, pp. 2425-2431, 2013.

[8] Q. Yu, X. Liu, A. Bouguettaya, and B. Medjahed, "Deploying and managing Web services: issues, solutions, and directions", The VLDB Journal—The International Journal on Very Large Data Bases, pp. 537-572, 2008.

[9] J. Hasan and B. El-Hajj-Diab, "BioFlow: A web-based declarative workflow language for Life Sciences", 2008 IEEE Congress on services-Part I. IEEE, 2008.

[10] T.R. Gruber, "Toward principles for the design of ontologies used for knowledge sharing?", International Journal of Human-Computer Studies, pp. 907-928, 1995.

[11] C. Hudelot, A. Jamal, and I. Bloch, "Fuzzy spatial relation ontology for image interpretation", Fuzzy Sets and Systems, pp. 1929-1951, 2008.

Development of Privacy-Preserving Online Monitoring Framework for Online healthcare Applications

Minsoo Kim and Youna Jung
Department of Computer and Information Sciences
Virginia Military Institute
Lexington, Virginia, United States
e-mail: kimm@vmi.edu, jungy@vmi.edu

***Abstract*— Privacy concern is one of the biggest obstacles in widespread adoption of online monitoring services on e-health applications, even though online monitoring can significantly improve the accuracy and quality of e-health applications. To prevent privacy loss during online monitoring, we have preliminarily proposed a privacy-preserving online monitoring framework (PPoM) that enables healthcare providers and patients to intuitively specify their own privacy policies and enforces patients' privacy policies in systematic manner during monitoring. For practical use of the PPoM framework, in this paper, we present the prototype development of the PPoM framework in detail. For the better understanding, we provide example usages from the standpoints of both healthcare providers and patients. To prove the performance of the developed prototype, we present evaluation results.**

Keywords- Privacy protection; online monitoring; framework

I. Introduction

Monitoring is one of the essential techniques to evaluate and enhance the performance of e-health applications by tracking and analyzing the online activities of patients, such as mouse clicks, frequency of use of an application, time spent in a particular page, media viewed, page navigation sequences, content entered into a textbox, location information, whether a mobile device is being used, etc.

Due to the sensitivity of the information that e-health applications often deal with, however, the protection of user privacy is critical. Control over the sharing of this information is of the utmost importance and urgency because indiscriminate monitoring, if inconsiderate of user privacy, may result in private health data being used for unwanted purposes and/or shared with unknown people [1][2][3]. It is therefore urgent and critical to enable the monitoring identifiable user data while protecting user privacy.

To this end, we have preliminarily proposed the PPoM framework [4] and the Health Insurance Portability and Accountability Act (HIPAA) Compliant Privacy Policy Language for e-health Applications [5]. The PPoM framework enables healthcare providers to collect necessary information without violation of patient's privacy preferences and HIPAA regulations and allows patients to enforce their own privacy preferences on the user side. To realize the proposed idea, we developed a prototype of the PPoM framework. The ultimate goals of the prototype development are as follows:

- For non-IT administrators of e-health applications
  - provide an intuitive way to describe privacy policies for their applications
  - provide an easy way to monitor patients' activities on their applications and collect patients' data without serious privacy breach
- For patients
  - provide a way to systematically verify an application's compliance with mutually agreed policies and HIPAA
  - Provide a way to rigorously protect their private data based on their preference on the user side

The rest of this paper is organized as follows. In Section II, our preliminary work is briefly introduced. In Section III, we present the architectural design of the PPoM framework and describe details about prototype development with example usages. In Section IV, we explain our test environment and present evaluation results. Our conclusion and future work are described in Section V.

II. Preliminary Work

To address the privacy issues on e-heath applications conducting online monitoring, the PPoM framework has been proposed [4]. In this section, we briefly introduce the overall architecture and the operational flow of the PPoM framework. As shown in Fig. 1, the PPoM framework consists of four components: the PPoM Service, the PPoM Browser, the PPoM Tools (PPoMT), and the HIPAA-Compliant Privacy Policy Language [5].

- *PPoM Service* – It is an online monitoring service that gathers only authorized user/usage data that users allow to monitor. By specifying user policies, patients can determine which data can be monitored. Then, the PPoM Service selectively collects user/usage data based on user policies. Unlike the existing monitoring services where user data are collected based on an application's policies and the policies are enforced by the application itself, the PPoM Service provides a way to refer user policies during online monitoring in a systematic manner rather than simply providing a written agreement.

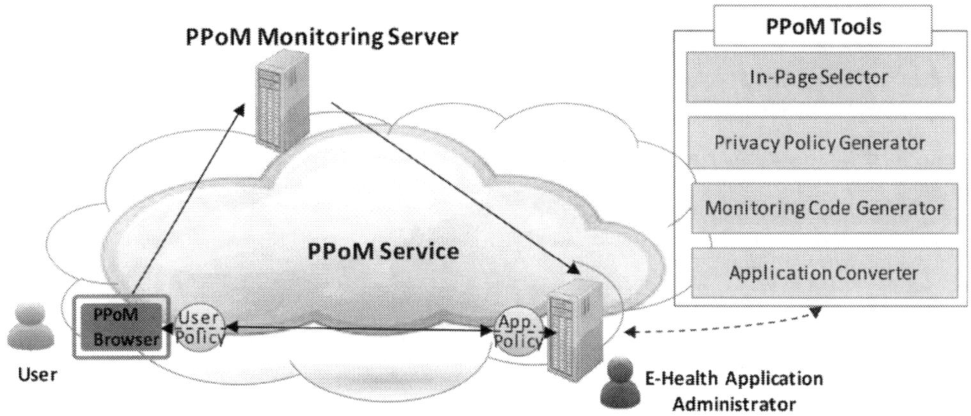

Figure 1. Overall Architecture of PPoM Framework.

- *PPoM Browser* – The privacy of patients must be protected even if a user is exposed to untrustworthy e-health applications that conduct indiscriminate monitoring in violation of user policies. Towards this end, the PPoM Browser understands a patient's privacy preferences, presents all user data being monitored, and protects a patient's privacy on the user side by blocking outgoing messages which contain data he/she does not want to disclose.
- *PPoMT* – Although patient monitoring is critical in e-health applications, it is difficult for healthcare providers to develop monitoring-enabled applications with application policies due to lack of professional IT knowledge. The PPoMT helps non-IT health professionals by enabling them to specify privacy policies for their applications, and to convert existing applications into privacy-preserving applications.
- *HIPAA-Compliant Privacy Policy Language* – Existing general-purpose privacy policy languages, including P3P [6], APPEL [7], and XPref [8], focus on generic user/usage data to be used for a variety of online applications and do not give careful consideration to health data. It is therefore impossible for both patients and e-health providers to precisely specify their privacy policies on health data in fine-grained level, and in turn, it lowers the performance of the PPoM framework. To address the lack of consideration of health data in existing privacy policy languages, the PPoM framework uses the HIPAA-compliant Privacy Policy Language employing the HIPAA profile [5], which allow an e-health provider to specify a privacy policy related to HIPAA regulations and enable a patient to specify his/her preference on health data in detail.

To use the PPoM framework, an online healthcare provider first needs to upload the source code or enter the URL(s) of his/her e-health application to the PPoMT and then select objects to be monitored and the corresponding privacy policies through the user-friendly interfaces generated by the In-page Selector. The Privacy Policy Generator then creates the application's policies by analyzing selected monitoring data and policies, while the Application Converter produces updated source code by inserting monitoring code generated by the Monitoring Code Generator into the original source code. The application policies and the updated source code must be deployed in the application server.

A patient now can use e-health applications without privacy concern. Whenever a patient enters a URL of e-health applications, his/her PPoM browser compares user policies and application policies. If they match, the application server sends PPoM-enabled pages which privacy-aware monitoring code is embed in. As the patient interacts with the application, the PPoM browser displays all user/usage data being monitored so that the patient can verify privacy protection during online monitoring. The monitoring code inserted in webpages collects only authorized user data based on user policies. The PPoM browser will block outgoing messages that violate the patient's policies.

III. Development of the PPoM Framework

To realize the proposed idea for privacy protection during online Monitoring, we developed a prototype of the PPoM framework. In this section, we describe details of implementation of each component. Note that the prototype of the PPoM framework is developed using mainly PHP and JavaScript. PHP is used for developing the backend of the PPoM Browser and the PPoM Tools while JavaScript including jQuery functions, HTML5, and CSS3 are mainly used for developing the user interfaces (UIs) on the client side. We use MySQL as a database of the PPoM service.

A. *PPoM Service*

The PPoM Service allows secure online monitoring on e-health applications. It provides privacy-aware monitoring APIs to online applications so that the administrators can embed the APIs in the webpages of their applications. Note that the privacy-aware APIs enables them to specify the type of data to be monitored, including types of health-related and

HIPAA-related data that are defined in the HIPAA Profile [5]. The APIs embedded in an e-health application collect data based on a patient's user policies, not an application's policies. It means that the privacy-aware APIs do not collect data if a patient prefers not to disclose. By doing so, the PPoM Service provides a way to protect a patient's privacy from indiscriminate monitoring.

Monitoring data collected contains general usage data such as *device category*, *operating systems*, *event*, and *time* and also user data including health-related data. All collected data are encoded in JavaScript Object Notation (JSON), a lightweight data interchange format, and sent by a patient's web browser to the PPoM Service server. The structure of a JSON monitoring data is shown in Fig. 2.

[ELEMENT_ID|ELEMENT_PATH] [EVENT_TYPE] [TIME] [DATA_ TYPE] [DATA] [DEVICE_INFORMATION]

- *ELEMENT ID*: It is a unique ID of a HTML element.
- *ELEMENT_PATH*: In case of dynamic webpages, a path from the root element is used as an ID if an element does not have ID. The path is unique for each element.
- *EVENT_TYPE*: It denotes that a type of an event occurred. The set of event types are as follows: {*entering a page*, *leaving a page*, *clicking an element*, *filling an element*}.
- *TIME*: It denotes the occurring time of an event
- *DATA_TYPE*: It is a type of monitoring data and must be specified based on the data types in the P3P data schema and the HIPAA Profile [5].
- *DATA*: It is the value of the monitoring data.
- *DEVICE_INFORMATION*: It includes a device's *category*, *operating system*, *language*, and *browser information*.

Figure 2. The structure of a JSON object for monitoring data

Let's assume that there is the *Blood Type* textbox, which its HTML code is shown in Fig. 3 (a). A patient enters "*A*" in the *Blood Type* textbox. The monitoring data on that textbox is then created as shown in Fig. 3 (b). At this time, the values of *EVENT*, *TIME*, and *DEVICE_INFORMATION* are automatically collected by JavaScript Built-in functions. Note that *blood type* is one of the health data type defined in the *Health* data schema of the HIPAA Profile. Before sending the monitoring data to the PPoM service server, a PPoM browser encodes the raw monitoring data as a JSON object as shown in Fig. 3 (c).

Fig. 4 illustrates the pseudo codes that enable the PPoM monitoring service on the server side (the PPoM Service server) and the application side (an e-health application). The `monitor` JavaScript function described in Fig. 4 (a) must be embedded in webpages of a PPoM-enabled e-health application prior to monitoring. When a target event is occurred, the `monitor` function captures and creates monitored data as a JSON object. Towards this goal, the function gathers necessary information by using JavaScript built-in functions and properties. Then, it invokes the `jQuery.ajax` function to communicate with the server-side scripts (the `receiveData` function shown in Fig. 4 (b)). The `jQuery.ajax` function converts a JSON object into a string and sends the string to the PPoM Service server through the HTTP POST method. When receiving a JSON string, the `receiveData` PHP module in the PPoM Service server converts the string into a JSON object and then stores monitoring data in the object in its database.

```
<input id="bloodtype" type="text" data-type
="health.bloodtype"/>
```

a) HTML Representation

```
ELEMENTID: bloodtype,
EVENT_TYPE: TEXTINPUT,
TIME: 2016-07-15T12:45:07,
DATA_TYPE: health.bloodtype,
DATA: Type A,
DEVICE_INFORMATION:DESKTOP(DEVICE CATEGORY),
WINDOWS(OS), ENGLISH(LANGUAGE),FIREFOX(BROWSER)
```

b) Raw Monitoring Data

```
{
 "ELEMENT_ID": "bloodtype",
 "EVENT_TYPE": "TEXTINPUT",
 "TIME": "2016-07-15T12:45:07"
 "DATA_TYPE": "health.bloodtype",
 "DATA": "Type A",
 "DEVEICE_INFORMATION":
  {    "DEVICE_CATEGORY": "DESKTOP",
       "OS": "WINDOWS",
       "LANGUAGE": "ENGLISH",
       "BROWSER": "FIREFOX"          }
}
```

c) JSON Object of the Raw Monitoring Data

Figure 3. Examples of monitoring data.

```
function monitor(object, event) {
   var monitoredData <= ID, EVENT_TYPE,
TIME, DATA_TYPE, DATA, DEVICE_INFORMATION
from the parameters object and event;
  jQuery.ajax({
    type: "post",
    url: "/PPoM/monitoring.php",
    data: JSON.stringify(monitoredData),
    contentType:"application/json; ",
    dataType: "json"                  });
}
```

a) JavaScript Function on the application side

```
function receiveData($monitoredData) {
  $object = json_decode($monitoredData);
  $link <= connection to database;
  $sql <= create INSERT query to store the
      monitored information ($object)
  mysqli_query($link, $sql);
}
```

b) PHP function in the PPoM service

Figure 4. Pseudo codes of the PPoM monitoring service on the application side (an e-health application) and the server side (the PPoM service).

B. *PPoM Browser*

The PPoM Browser provides three ways to protect user privacy. First, it enables a non-IT patient to intuitively specify his/her privacy preferences. Through its user-friendly interfaces shown in Fig. 5, a patient can define three different levels of policies, the General Policies (GP), the Application-specific Policies (AP), and the Page-specific Policies (PP). First, the General Policies describe a patient's general preference regarding data sharing. An Application-specific Policy is applied to an online application and it affects across webpages in an application while a Page-specific Policy is applied to a particular webpage of an application. To define a GP, a patient needs to specify data types that a patient allows or disallows to be monitored across different applications. To specify a AP or a PP, a patient needs to enter the url of an application or a webpage. If two or more policies conflict, then the most specific policy takes precedence. Note that the prototype of the PPoM Browser we developed currently supports the Application-specific policies and the Page-specific polices.

Figure 5. A screenshot of the user interface of the PPoM browser that displays all data being monitored when a user turns on the privacy-preserving mode.

Second, it displays all data being monitored when a patient uses an e-health application. Although a patient agrees to an application's policies, it is critical to verify the application's compliance with the mutually agreed policies. By turning on the privacy-preserving mode of the PPoM browser, a patient can easily figure out that what usage/user data are being monitored and protect data from unwanted monitoring without any IT knowledge and skills. An example use of the developed PPoM browser is shown in Fig. 5. The browser displays which data is being monitored and who is the recipient of the data by using different-colored check marks. The *red* check marks mean that the data is protected and there is no recipient. The *orange* check marks mean that only the first party (for example, an e-health application that a patient is using) is receiving the data. The *green* check marks mean that third parties (for examples, other healthcare providers referred by the first party, advertisement companies, and payment companies) are also receiving monitoring data. A summary of monitoring data, including not only general usage data (such as *HTTP* data, *Clientevents*, and *Cookies* types of *Dynamic* data schema in P3P) and user data (*name*, *bday*, and *location* of *User* data schema in P3P) but also health-related data (*blood type*, *disabilities*, and *disease history* of *Health* data schema in the *HIPAA profile*) on a webpage is displayed in the status bar.

Third, it enables a patient to stop monitoring on a specific page or web object if he/she finds out fraud activities that are against the mutually agreed policies. Towards this end, the PPoM Browser first renders all clickable web objects -- such as buttons and objects handled by JavaScript click-event handler-- and input HTML elements, such as textbox and checkbox. It then creates checkboxes for each of them. By changing the colors of checkmarks, a patient can easily control data sharing. At this time, a patient has multiple options for selecting objects: 1) select all clickable and input elements, 2) select all clickable elements, 3) select all input elements, 4) select an individual (clickable or input) element, or 5) select *None*.

To block monitoring, a PPoM browser needs to generate and run JavaScript codes based on a patient's selection in real-time. Before explaining the blocking process further, let's assume that the main functionalities of an e-health application does not depend on JavaScript. To block online monitoring on a particular web element, a PPoM browser disables JavaScript event handlers that are associated with the selected web elements, and in turn, a monitoring JavaScript using those handlers will be disabled. For example, if a patient marks the *Blood Type* textbox in *red* color to block monitoring on that textbox, then the following JavaScript code is generated to disable the PPoM monitoring JavaScript on the *Blood Type* textbox (The ID of the textbox element is "BLOODTYPETXT"): `$("body").off("keyup keypress change click blur", "#BLOODTYPETXT")`. The generated code then removes five event handlers from 'BLOODTYPETXT' element by invoking jQuery `off` function. When the blocking code runs, none of monitoring services obtain data from that textbox.

C. *PPoM Tools (PPoMT)*

The PPoMT is a toolkit on the server side, which enables non-IT healthcare providers to generate application policies and monitoring codes for their own e-health applications through intuitive user interfaces and also easily upgrade their existing applications into a PPoM-enabled e-health application. The detailed explanation for each component in the PPoMT is below.

1) In-Page Selector

The goal of the In-Page Selector is to generate modified webpages that shows selectable HTML elements on webpages so that an administrator of an e-health application can select usage/user data to be monitored and policies corresponding to each data, each page, or an entire application. When an administrator selects a web element or a group of elements, another window is popped up to specify a privacy policy about the selected element(s). As shown in the Fig. 6, a non-IT healthcare provider can create a P3P-based application policy which deals with health data and HIPAA regulations. Towards this end, a set of data and policies that are selected by an administrator is delivered to the Privacy Policy Generator and the Monitoring Code Generator.

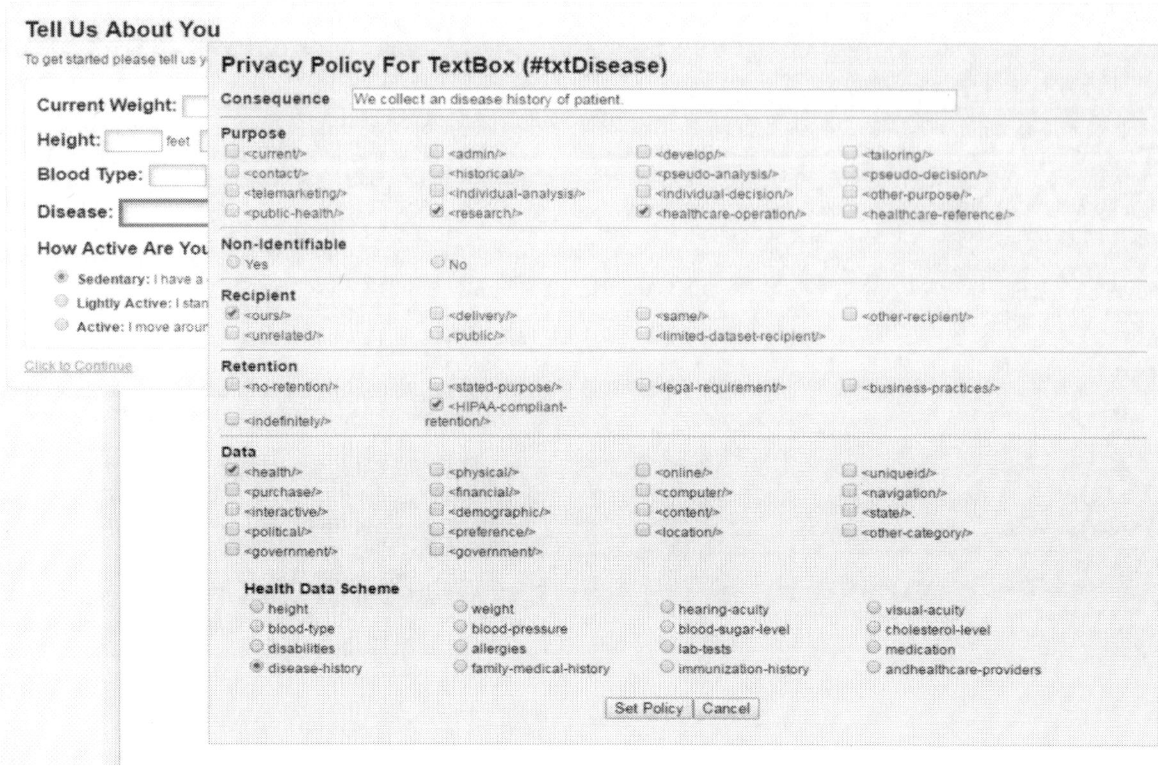

Figure 6. An example of the screenshot of the PPoMT.

2) Privacy Policy Generator

The Privacy Policy Generator generates an application's policies in the HIPAA-Compliant Privacy Policy Language [5] based on an administrator's selection. Not only an administrator of an e-health application but also a patient can use the Privacy Policy Generator to specify a patient's user policy for a specific e-health application. To do so, a patient needs to enter the url of an e-health application and select *User Policy Generation* rather than *Application Policy Generation*. If the *User Policy Generation* is selected, the Policy Generator creates APPEL policies employing the HIPAA profile.

3) Monitoring Code Generator

When receiving a set of data to be monitored and relevant policies, the Monitoring Code Generator produces privacy-aware monitoring codes for an e-health application. Toward this goal, it first checks if each selected element has an ID. If an element doesn't have an ID, it assigns a unique element ID and generates JavaScript code using the assigned ID. Depending on the type of an application, a *static* application or a *dynamic* application, a generated element ID will be different. If an e-health application is a *static* application that the same HTML code stored in an application's server is delivered to all patients' browsers, the Monitoring Code Generator assigns an absolute ID to an element. Let's assume that a webpage has several textboxes and its HTML code is shown in Fig. 7 (a). In this example, the *Current Weight* textbox has its ID (`"WEIGHT"`) but other three textboxes -- the *feet*, the *inches*, and the *Blood Type* textboxes-- do not have their IDs.

If the webpage is a *static* page, then the Monitoring Code Generator automatically creates IDs for three textboxes: `"PPOM-ELEMENT-0001"` and `"PPOM-ELEMENT-0002"` for the *feet* and *inches* of the *Height* textbox and `"PPOM-ELEMENT-0003"` for the *Blood Type* textbox. The generated monitoring code for the *static* webpage is shown in Fig. 7 (b). On the other hand, if an application is a *dynamic* application that the HTML codes are dynamically generated by an application server on demand, a path of an element from a root of a Document Object Model (DOM) object is used as a unique ID because an element's path is unchangeable. The monitoring code for a *dynamic* webpage is shown in Fig. 7 (c). As you can see, except the pre-defined ID of the *Current Weight* textbox, for other three textboxes which do not have IDs, their paths are used to identify each textbox. For examples, `"input[type=text]:nth-of-type(2)"` and `"input[type=text]:nth-of-type(3)"` for the *feet* and *inches* of the *Height* textbox and `"input[type='text']:nth-of-type(4)"` for the *Blood Type* textbox. Note that the PPoMT uses the PPoM Service so privacy-aware monitoring script code is generated as default, but it is possible to use different monitoring services such as Google Analytics.

```
<body>
Current Weight:<input id="WEIGHT" type="text">lbs.
Height: <input type="text">feet
        <input type="text">inches
Blood Type: <input type="text">
<input id="BUTTON1" type="submit" value="Submit">
</body>
```

a) A partial HTML code of an e-health application shown in Fig. 5

```
<body>
Current Weight: <input id="WEIGHT" type="text">
          lbs.
Height:<input id="PPOM-ELEMENT-0001" type="text">
          feet
      <input id="PPOM-ELEMENT-0002" type="text">
          inches
Blood Type:
      <input id="PPOM-ELEMENT-0003" type="text">
<input id="button1" type="submit" value="Submit">
</body>

<script>
   $("#WEIGHT").change(function() {
      monitor($(this), "change");
   });
   $("#PPOM-ELEMENT-0001").change(function() {
      monitor($(this), "change");
   });
   $("#PPOM-ELEMENT-0002").change(function() {
      monitor($(this), "change");
   });
   $("#PPOM-ELEMENT-0003").change(function() {
      monitor($(this), "change");
   });
   $("#BUTTON1").click(function() {
      monitor($(this), "click");
   });
</script>
```

b) HTML code converted by the PPoMT in case of a static application

```
<script>
$("#WEIGHT").change(function(){
      monitor($(this), "change");});
$("input[type='text']:nth-of-
type(2)").change(function()
   { monitor($(this), "change");});
$("input[type='text']:nth-of-
type(3)").change(function()
  { monitor($(this), "change");});
$("input[type='text']:nth-of-
type(4)").change(function()
  { monitor($(this), "change"); });
$("#BUTTON1").click(function(){
      monitor($(this), "click"); });
</script>
```

c) Monitoring JavaScript code generated in case of a dynamic application

Figure 7. Examples of monitoring code generated by the PPoMT.

4) Application Converter

This component produces a PPoM-enabled application by inserting monitoring codes generated by the Monitoring Code Generator into DOM objects for each webpage in an application. Depending on a type of an application, the conversion process would be different. In case of a *static* application, the Application Converter can systematically generate the updated HTML code. If an application is however a *dynamic* application, the PPoMT provides monitoring script code only and then an administrator should manually insert the generated monitoring code into the server-side program that dynamically creates webpages.

IV. Evaluation Results

To test the usability of the proposed PPoM framework, we first develop two types of sample e-health applications that each has ten webpages containing different numbers of monitoring elements without IDs and a prototype of the PPoM framework, including the PPoM service, the PPoM browser, and the PPoMT. Then, we test the performance of the PPoMT according to the evaluation plans described in [4].

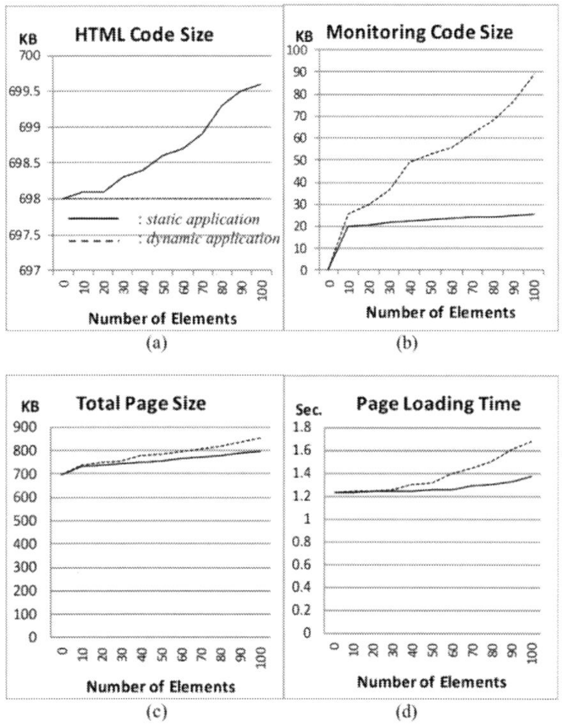

Figure 8. Evaluation Results.

As shown in Fig. 8, the size of HTML code that generated by the Application Converter for *dynamic* webpages is zero since the PPoMT will not generate HTML code for a dynamic application while the size of HTML code for *static* webpages increases linearly as the number of the monitoring elements increases.

In case of *static* webpages, the size of the generated monitoring code remains steady once it reaches a certain size even though the number of monitoring elements increases linearly. However, the size of monitoring code for *dynamic* webpages linearly increases according to increase of the number of monitoring elements (See Fig. 8 (b)). This is because specifying paths from root is costly, especially for complex web pages.

As you can see in Fig. 8 (c) and (d), in case of *static* webpages, the number of monitoring elements does not affect the size of converted webpages that produced by the PPoMT and also the page loading time. However, in case of *dynamic* webpages, the increase in the number of monitoring elements affects the page loading time. If a *dynamic* page is complex, the loading delay becomes a big obstacle. It is one of our challenges to find out a way to minimize the loading delay caused on dynamic webpages.

TABLE I. FAILURE RATIO IN THE PPOM SERVICE

	Monitored	Not Monitored
Allowed	(a) 4,897	(b) 345
Not Allowed	(c) 0	(d) 3,477

TABLE II. SUCCESSFUL BLOCKADE RATIO IN THE PPOM BROWSER

	Sent	Blocked
Allowed	(e) 5,242	(f) 0
Not Allowed	(g) 0	(h) 3,477

To evaluate the privacy protection of the PPoM Browser and the PPoM Service, we produce five hundreds of different sets of patients' privacy preferences (i.e., allow or disallow monitoring on particular web elements) and activities on a sample static application (i.e., navigating webpages, clicking buttons, or entering data in input elements). Using the sets of user policies and activities, we test the PPoM browser and the PPoM Service. To evaluate the privacy protection on the prototype of the PPoM framework, we measure two factors: the failure ratio ($c/(a+c)$ in Table I) and the successful blockade ratio ($h/(g+h)$ in Table II).

The failure ratio evaluates privacy protection on the server side (the PPoM Service server) and the successful blockade ratio evaluates protection on the client side (the PPoM Browser). As shown in Tables I and II, none of user/usage data that patients do not allow to be monitored was captured by the PPoM Service and none of the unauthorized data was sent from the PPoM Browser. However, as shown in Table II, we found some data loss in the PPoM Service server. The PPoM Browser sent 5,242 data ($a+b$), but the PPoM Server received only 4,897 data (a). It may be caused by heavy load of transaction or overheads for checking application and user policies. To figure out the source of the resulting data loss, we need to investigate more in the future.

V. CONCLUSION

Privacy protection is critical for widespread use of e-health applications. Without proper methods for the privacy preservation, people may keep hesitating to use e-health applications, even though e-health applications help them access healthcare services in easy and convenient way at the reduced cost. To address the privacy issue, we develop the prototype of the PPoM framework that protects user privacy in both the application side and the user side. To prove the performance and the usability of the developed prototype, we present the evaluation results. Towards our ultimate goal, however, we need to complete the following tasks in the future:

- Research a way to reduce data loss ratio on the PPoM service.
- Investigate a method to reduce the size of the monitoring codes generated by the PPoMT, especially for dynamic e-health applications.
- Field tests of the prototype with actual e-health applications.

REFERENCES

[1] J. R. Mayer and J. C. Mitchell, "Third-Party Web Tracking: Policy and Technology," Proc. IEEE Symp. on Security and Privacy (SP '12), IEEE Press, pp. 413-427, 2012.

[2] M. Bilenko and M. Richardson, "Predictive client-side profiles for personalized advertising," Proc. ACM SIGKDD conf. on Knowledge discovery and data mining (KDD '11), ACM New York, pp. 413-421, 2011.

[3] A. McDonald and L. F. Cranor, "Beliefs and Behaviors: Internet Users' Understanding of Behavioral Advertising," TPRC 2010 Social Science Research Network (SSRN), August 16, pp. 1-31, 2010, Available from http://ssrn.com/abstract=1989092 [retrieved: September, 2016].

[4] Y. Jung, "Toward Usable and Trustworthy Online Monitoring on e-health Applications," International Journal On Advances in Life Sciences, vol. 8, numbers 1 and 2, pp. 122-132, June, 2016

[5] Y. Jung and M. Kim, "HIPAA-Compliant Privacy Policy Language for e-health Applications", Proc. the 6th International Conference on Current and Future Trends of Information and Communication Technologies in Healthcare, September, London, United Kingdom, in press.

[6] P3P 1.1. http://www.w3.org/TR/P3P11/ [retrieved: September, 2016].

[7] A P3P Preference Exchange Language (APPEL) version 1.0. https://www.w3.org/TR/P3P-preferences/ [retrieved: September, 2016].

[8] R. Agrawal, J. Kiernan, R. Srikant, and Y. Xu, "An XPath-based preference language for P3P," Proc. the 12th international conference on World Wide Web (WWW '03), ACM, New York, pp. 629-639, 2003.

AUTHOR INDEX